CORPORATE STRATEGY
FOR DRAMATIC PRODUCTIVITY SURGE

CORPORATE STRATEGY FOR DRAMATIC PRODUCTIVITY SURGE

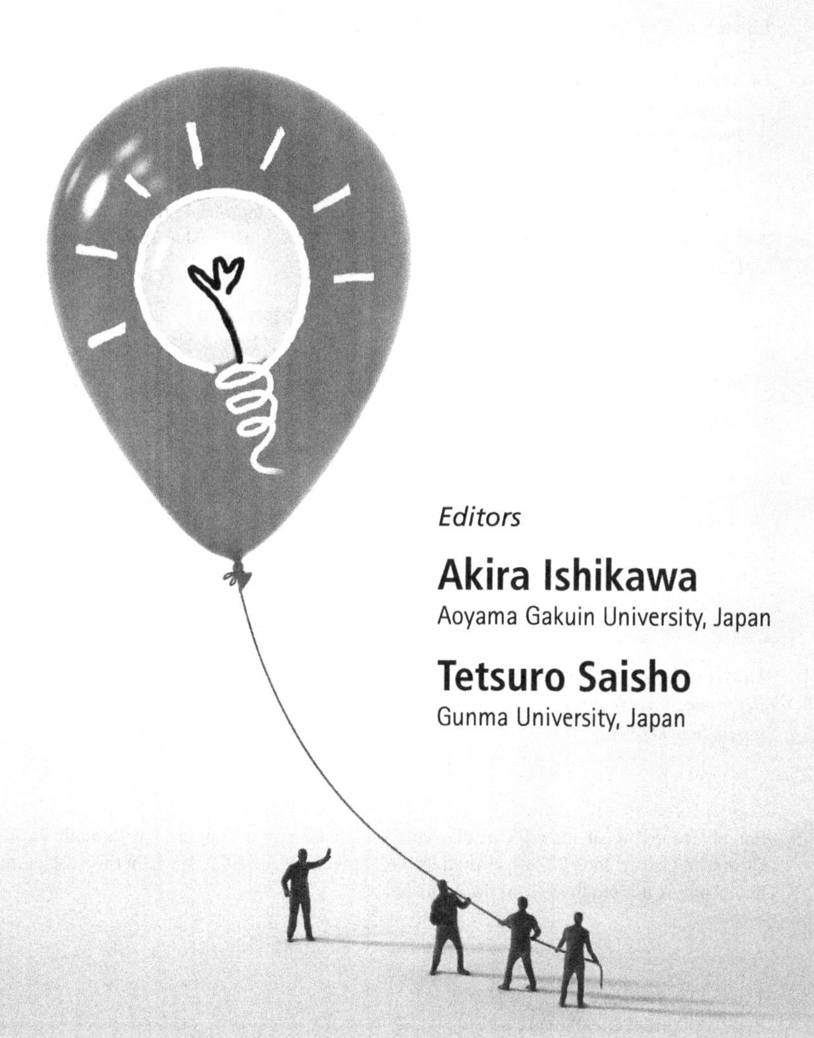

Editors

Akira Ishikawa
Aoyama Gakuin University, Japan

Tetsuro Saisho
Gunma University, Japan

World Scientific

NEW JERSEY • LONDON • SINGAPORE • BEIJING • SHANGHAI • HONG KONG • TAIPEI • CHENNAI

Published by

World Scientific Publishing Co. Pte. Ltd.
5 Toh Tuck Link, Singapore 596224
USA office: 27 Warren Street, Suite 401-402, Hackensack, NJ 07601
UK office: 57 Shelton Street, Covent Garden, London WC2H 9HE

Library of Congress Cataloging-in-Publication Data
Corporate strategy for dramatic productivity surge / edited by Akira Ishikawa, Aoyama Gakuin University, Japan & University of Hawaii & Tetsuro Saisho, Gunma University.
pages cm
Includes index.
ISBN-13: 978-9814449298 (hardcover : alk. paper)
ISBN-10: 9814449296 (hardcover : alk. paper)
1. Organizational effectiveness. 2. Strategic planning. 3. Industrial productivity.
4. Reengineering (Management) I. Ishikawa, Akira, 1934– II. Saisho, Tetsuro.
HD58.9.C668 2013
658.4'012--dc23
2012046039

British Library Cataloguing-in-Publication Data
A catalogue record for this book is available from the British Library.

Typeset by Stallion Press
Email: enquiries@stallionpress.com

Printed in Singapore.

Preface

In the mid 1990s, the book *Reengineering the Corporation: A Manifesto for Business Revolution*, authored by the former MIT professor, Michael Hammer, and the business consultant, James Champy, made waves, giving rise to the buzzwords, "business process re-engineering", or simply "re-engineering".

Amid such a time, I took part in a seminar and panelist discussion held at the Nikkei Hall on April 2nd, 1994. The event was titled, "Re-engineering Revolution — New Business Approaches and Information Technologies for Escaping the Depression and Surviving the 21st Century" and I presented at that time the idea of "a digit-difference effect, or an effect that is more than 10 times larger in capacities and smaller in costs." (henceforth referred to as a productivity surge/super effect) (Nikkei Information Strategy, June 1994, page 31).

What these words mean is this: To radically and dramatically improve an organization, which is highly specialized and involves processes that is suitable to be compartmentalized, you must review fundamental organizational or business policies, rules, and protocols that have become firmly established. Only then, can you attempt to redesign an organization, lines of duty, business processes, management, and information systems, keeping in mind the multilateral perspective of business processes.

My proposal aimed not only to make improvements but also maximize value for both customers and stakeholders. At one time, people were vociferous about the concept of regeneration, but since this mode of thinking had reached an impasse, the question of how to specifically achieve regeneration was urgently needed.

Furthermore, this book has introduced not only a US car company actively adopting IT, but also a case that shows how a Japanese car company radically restructured its accounting division by analyzing its business operations. Consequently, the company went on to reduce more than 500 of its personnel and dramatically streamlined and sped up business operations. Since the achievement of such a result is believed to be impossible by the adoption of IT alone, it was only natural that Japanese companies and executives became interested to know how it was made possible.

When we re-examine these cases or the concept of re-engineering from the point of view of productivity, we can see that a dramatic improvement in the relationship between input and output. However, this viewpoint is most importantly because it pertains to a multilateral approach that seeks out radical changes in individual processes and the whole framework of processes.

Yet, since then, the thought process and the spirit behind re-engineering appears to have died out, and in the survey conducted by Switzerland's IMD on international competition in fifty-five countries and regions across the world, an assessment on Japan's productivity in the category of business efficiency has dropped to 42nd place (according to the 2007 survey). Japan now ranks so low that it has become quicker to spot its place in the ranking by searching from the bottom of the list. The editor and the writers of this book were adamant that such a situation must not be ignored.

Thus, this book confronts the question of whether Japanese thinking behind productivity and related activities are in a state of decline and attempts to answer it by introducing remarkable, dramatic cases based on research.

The starting point of "dramatic" or "radical" productivity, as Hammer and Champy frequently mention, should be the productivity surge — that is, the effect of achieving "over a tenfold increase in performance for the same cost" or "lowering the cost to one-tenth its original level while maintaining the same standards, indicators, and conditions".

To show such an effect, eight writers have researched past scenarios and compiled them into around four to eight cases, creating a total of more than fifty presentable accounts. Their extent and scale go beyond the scope of the private enterprise, covering not only the global, national, and societal, but also public enterprises. Additionally, among private enterprises, large enterprises as well as medium- and small-sized businesses are covered.

In American finance-oriented capitalism, maximization of profit and minimization of cost tend to be immediate goals, but in cases where great importance is attached to productivity surge, the focus was on avoiding loss of a pluralistic, multiplex, multifaceted orientation by incorporating functional aspects directly related to productivity, such as time reduction, speed increase, expansion of capacity, improved sensitivity, improved accuracy, increase in effectiveness and efficiency, change in size, and discovery of new materials.

The more these strategies developed, the more they improved dramatically, leading to several thousands-to, tens of thousands-fold improvements in effectiveness, efficiency and surprising cost reductions. However, in some cases, the cumulative effect of the productivity surge was not always positive, but also negative. This is because risk factors were present in those case examples, and the tragedies began when those risks turned out to be "unexpected factors".

Each part of the book is put together based on the type of strategy involved, so the reader is free to begin reading about any strategy that interests them.

This book can be read by administrators, researchers and employees working in government organizations, companies, NGOs, NPOs, and college students, rather than readers whose interests lie in science and technology or liberal arts. In the interests of helping readers grasp the subjects more easily, the scope and range of sections classify subjects by the categories of world, nation, society, and enterprise.

Suggestions, comments, or opinions from readers and other thoughtful individuals are most welcome.

Akira Ishikawa
Co-editor

Contents

Introduction

Tetsuro Saisho

With the development and spread of information and communication technologies, we have in our modern society an environment for the workings of an information-intensive society in which any necessary information can be effectively accessed anytime, anywhere and by anyone, again and again. These include not only large enterprises, but also small- and medium-sized businesses, non-profit institutions, educational institutions, research organizations, administrative bodies, and even individuals, who make uses of the information to advance or facilitate organizational activities or to enhance personal lives.

Such a post-industrial information-intensive society is a society whereby activities of organizations and individuals can be carried out in an egalitarian, fair, and impartial manner. This is made possible by the fact that such a society transcends the constraints of time, space and conceptual differences that arise between different nations, regions, political factions, economic theories, management styles, societies, cultures, philosophies, ethical standards, and views on education, through self-propagrating open networks that continue to expand borderlessly and globally.

The changes in the activities of organizations and individuals brought about by the advent of the information-intensive society are expanding, as the degree of their impact grows. For example, the activities of organizations and individuals were largely affected when information of all sorts (including the news, weather forecasts, novels, films, songs, theatrical shows, sports, games, animations, shopping data, travel information, advertisements, and publicity campaigns) became available in digitized formats.

There was a fundamental shift in the nature of information. Formerly, bodies of information, such as letters and figures, voice recordings, still images, animations, pictures, were things that could not be "processed" at will (analog) by a human being. However, by turning them into something (digital) that can be "processed" — visualized objects with appearance, phenomena, and relationships that can be numerically expressed as data — it has become possible to handle information in various fields.

In an environment capable of handling such digitized information, productivity surge can be achieved more easily in these areas:

- Operational strategies, such as time reductions, speed boosts, and expansion of capabilities;
- Global-scale operations, such as improvements in sensitivity, precision, and boosts in effectiveness and efficiency;
- Market strategies, such as size changes, cost reductions, and sales expansions;
- Functional aspects directly related to productivity, such as quantitative increases and discovery of new materials.

This is because digitized information can be handled as an integrated unit regardless of physical distance and location, enabling its users to take optimal advantage of the conveniences of the information itself and bring about productivity surges in new products and businesses. The conveniences of information include quality retention, multichannel nature, ease of processing, high-speed processing of transmissions, swiftness of retrieval, the real-time aspect of transmissions, interactivity, and asymmetry.

Open networks supporting the modern information-intensive society, such as the Internet, are like spider webs linking countries, regions, localities, areas, and sites found across the world. An open network is a widely recognized information network that is indispensable to a society's infrastructure, and that is accessible even if parts of it were cut off because of the availability of alternative routes.

In an open network, it is possible, through effective uses of information and communication technologies, to share, integrate and standardize various types of information, in a way that large numbers of new businesses and services are able to offer conveniences and benefits that can be enjoyed.

In addition, in an information-intensive society, it has become possible for anyone to enjoy the conveniences and benefits on an equal basis.

In this kind of society, in contrast to a brick-and-mortar one, its members rapidly form large societies, markets, and communities. Furthermore, this kind of society involves not only large enterprises, but also small- and medium-sized businesses, non-profit organizations, educational institutions, research organizations, administrative bodies, and even individuals. Whether they do so consciously or not is a separate matter.

In this book, we cover contents that concretize productivity surges impacting new products and businesses, by introducing case examples of *Twitter* use, web conferences, multihead weighers (known as computer scales in Japan), thin clients, algorithms, data mining, e-learning, common points, electronic money, and ID links. All illustrated cases of productivity surges in this book exhibit the characteristics of societal information mentioned above.

Incidentally, productivity surges have appeared even in the manufacturing industry, a sector not usually considered to be directly related to an information-intensive society. This is the case not only for large enterprises, but also medium- and small-sized firms and venture companies.

For example, in covering the manufacturing industry, we introduce case examples of business models that involve building new structures, reinventing technologies of other fields, creating models for product development, changing the learning processes applied in product development, reusing resources by making conceptual changes, introducing lump packaging instead of packing products separately, and developing new technologies and applying them to business operations.

To illustrate the productivity surges distinctive to the manufacturing industry that impact new products and new businesses, we introduce case examples of the following companies:

- Large companies — *Hitachi, Panasonic, Toshiba, Fuji Electric, Sharp, Toray, Mitsubishi Heavy Industries*, and *Tsubakimoto Chain*; the foreign firms, *P&G, IBM, GE, Dell*;
- Small and medium enterprises— *Gunkyo Factory*, *Japan Highcomm*, *Anzai Manufacturing*, *Toko Electric Corporation*, *Ishida*, and *Hikari Kogyo* (currently *Art-Hikari*).

In addition to the manufacturing industry, which a traditional industry, productivity surges have also been appearing in the distribution, retail, warehouse, railway, aviation, and consulting sectors, impacting new products and new businesses in those sectors.

To illustrate the productivity surges distinctive to the distribution and retail sectors that impact new products and new businesses, we introduce case examples of the following companies: *Eisai Distribution*, *Yamada Denki*, *Amazon.com*, *Japan Imagination*, *Zappos*, *Bals*, *Mitsui Soko*, *Kyushu Railway Company*, *Ryanair*, and *Nomura Research Institute*.

For businesses in the distribution and retail sectors, times have become such that unless they can make drastic proposals *vis-à-vis* their competitors — proposing things such as restructuring business processes or making new uses of existing technologies — productivity surges cannot be achieved. This is because, under the progress of the information-intensive society, information acquisition, volume, and the speed of propagation have all increased to the extent that comparisons to the past can no longer be made.

In addition, rival and competing firms quickly take notice of the new ideas that propel productivity surge and analyse and evaluate them. When product patents (for those usually seen in the manufacturing industry) or business model patents are not granted, imitations and forgeries ensue within a short period of time and any competitive edge brought about by a productivity surge disappears .

Asia-related businesses in particular have a low level of awareness for intellectual property rights. In fact,there is a remarkable trend for them to infringe on the four major intellectual property rights: patent rights, model utility rights, design rights, and trademark rights.

What must be assumed here is that amid the progress of the information-intensive society, rival and competing firms, are just like your firm in terms of seeing the rise in information acquisition, volume, and speed of propagation. In other words, this means that in the distribution and retail industries, for example, the *modi operandi* that appear to be successful at first glance — thanks to existing productivity surges — are not substantial enough to serve as absolute strategies for companies to carry out permanent activities, to grow, and to realize lasting progress.

For this reason, to achieve dramatic productivity surges, it is utterly inconceivable to rely simply on fleeting thoughts, flashes of inspiration, intuition, or even on ill-prepared measures taken to deal with immediate issues. To achieve dramatic productivity surges, it is necessary to build detailed business plans and programs and carry them out one by one and accumulate experience, while practicing adequate, selective, and concentrated allocation of management resources to pave the way towards realizing dramatic productivity surges for your company.

As we have seen until now, with the advancement of the information-intensive society and the subsequent impact of various environmental changes that affect not only large enterprises, but also small- and medium-sized businesses, non-profit organizations, educational institutions, research organizations, and administrative bodies, we have been stating that productivity surges can serve as the sources of competitive edges — functional competitive edges directly related to productivity, such as those arising from time reductions, speed boosts, expansion of capabilities, improvement in sensitivity, improvement in accuracy, boosts in effectiveness and efficiency, size changes, and discovery of new materials.

Whether or not your firm actively utilizes information and communication technology, you can pave the way to bring about outstanding strategies through by realizing productivity surges. These include developing new technologies formerly believed to be impossible, building new structures, redeploying technologies designed for other fields, building product development models, or changing the learning processes in product development,. Suffice to say, the more these strategies are in sync with these productivity surges, the more you will see dramatic improvements at a holistic level. This in turn leads to surprising cost reductions and remarkable improvements in effectiveness and efficiency — improvements that are several thousand or even tens of thousand-fold.

On the other hand, there are various wastes, irregularities, and impracticalities in the course of running a business. This applies not only to large enterprises, but also small- and medium-sized businesses, non-profit organizations, educational institutions, research organizations, and administrative bodies. For example, wastes, irregularities, and impracticalities of management caused by large-scale reductions in new businesses or their withdrawal; changes in an industry or business conditions; or a failure to launch a company, have a large impact on organizations and the damage to an organization running a business could be immense.

Furthermore, consumer tastes in the present day, are diversified. Thus, requirements for products and services are also exacting. As a result, companies develop mostly unsellable new products or services; or make unreasonable and needless capital investments (including new and additional ones) and consequently fail to make any effective use of facilities, incurring unnecessary expenses that accompany considerable wastes, irregularities, and impracticalities.

The amount of a single unnecessary expense will more or less affect an organization. A single unnecessary expense may not be substantial enough to cause direct fatal harm to an organization (depending on its financial health). However, when the frequency of such an occurance rises considerably, the total amount of unnecessary expenses becomes substantial. In this case, there is a negative productivity surge, which demonstrates that the aggregate of unnecessary, minor expenses could lead to a significant undermining of an organization's financial health.

In modern society, measures considered to be suitable for the purpose of improving business operations in general, in many cases, turn out to be those that involve the elimination of wastes, irregularities, and impracticalities. These include overproduction arising from work operations or inventory management, cessation of machines, mechanical malfunctions, stand-by management, operational checks, inspections, appraisals, production of defective units, product processing, and oversight or neglect in operating instructions and manuals.

While these wastes, irregularities, and impracticalities may seem small or negligible in comparison with the above-mentioned unnecessary expenses, these small wastes, irregularities, and impracticalities are inherent in the developments of all organizations and individuals and may therefore accumulate to a massive scale. Consequently, it would not be an exaggeration to say that the total amount of unnecessary expenses incurred through wastes, irregularities, and impracticalities will ultimately influence the survival of an organization.

Thus, with regard to the improvement of the business operations, companies are carrying out activities that aim to eliminate wastes, irregularities, and impracticalities found in general business processes on a daily basis. Reforms are always being carried out to take into account the actions of all members of the organization at the workplace, while working on standard procedures to improve and sustain the effect of the reforms at an individual level.

PART I

Case Studies of Productivity Surges in Management Functions

Section 1: Time Reduction

1

Welding Steel Plates Six Times Faster

Tetsuro Saisho

The Development of a New Technology

The case introduced here illustrates a drastic time reduction achieved through the development of a new technology. It is an account of the improvement in the speed of welding steel plates and the development of a welding machine that enhances the hardness of steel plates. It is also an account of the application of this improvement and development. The newly developed welding machine is capable of welding a steel plate six times faster than the conventional processing equipment.

Until now, in automotive factories, when adopting high-powered welding techniques for welding steel plates, the higher the intensity applied, the bulkier and heavier the plates became. In addition, the weight of the welding equipment was very heavy as well, making its usage inefficient since it was immovable and therefore could only be used in one place as a stand-alone unit. It was impossible to use it in combination with other machine tools.

In contrast, the new welding machine was movable and capable of being applied for a wide range of uses, since its weight was one-seventh its predecessor. It became possible to consider multi-faceted applications for the new welding machine. Furthermore, since it has also become possible to maintain a fixed intensity for even thin steel plates, engineers producing automobiles using these steel plates were able to drastically lighten the completed bodies of the cars, a factor that led to improved fuel efficiency.

For companies that carry out steel plate welding at an automotive factory, the electrical passage necessary until now to fuse steel plates together with conventional equipment was 40 mm/second. However, with the new models, this task could be performed with an electrical passage of only 6 mm/second. Due to this, the new machines produce lesser irregularities in the composition face even if the operator were to weld by moving the electrodes quickly.

Automobile parts-makers, generally adopt spot welding, which entails welding in the manner of pricking multiple steel plates with needles. The new welding machine joins by using a technique known as seam welding. In effect, this technique joins by fastening two to three steel sheets with a ring-shaped electrode and draining the electricity before sewing.

The new model using this technique of seam welding, by raising electrical efficiency, has down-scaled the power source and power transformer and saved the weight of the newly developed welding machine of around 140 kg at once. It is also attached to welding robots in automotive factories, adjusted to levels near those of spot welders of approximately 100 kg. The speed of welding, compared to spot welding, increases by six times with this model, leading to increased efficiency in assembly operations.

In such a situation, *Hikari Kogyo* Co., Ltd. (currently *Art-Hikari*, Co. Ltd., based in Tatebayashi city, Gunma, President Kazutoshi Furukawa), in addition to boosting the speed of processing steel plates with seam welding, has developed a welding machine that has made it possible to raise the hardness of welded steel plates[1].

A salient feature of the seam welding technique is that it heightens the unit strength by a large margin to weld by line instead of by point, which was how it had been conventionally welded with in spot welding technique. At conventional welding sites, seam welding has been used in nuclear-reactor processing, which requires high-performance welding. However, since the welding equipment itself weighs around one ton, this technique did not become widespread in automotive factories.

Nonetheless, anticipation has risen for the application of seam welding in various fields, thanks to the weight reduction of welding equipment and the weight reduction and hardening of welded steel.

Seam Welding and Spot Welding

Primarily, welding is an operation that fuses two or more metallic components by heating them and adding pressure to fuse their junctions to ensure continuity at the molecular level. Even in the case of typical welding

<table>
<tr><td rowspan="6">Fusion weld</td><td>Gas welding</td><td colspan="2">Oxy-acetylene welding (mixed gas welding of oxygen and acetylene)</td></tr>
<tr><td rowspan="2">Arc welding</td><td>Non-consumable electrode type</td><td>Tungsten inert-gas welding (aka TIG arc welding)
Plasma welding (welding via plasma arc use)</td></tr>
<tr><td>Consumable electrode type</td><td>Arc welding with covered electrode (stick welding, hand welding)
Submerged arc welding (uses flux, welding wire)
MIG welding (uses only an inert gas as shielding gas)
Carbon dioxide gas-shielded arc welding (uses carbon dioxide as shielding gas)
Self-shielded arc welding (does not externally supply shielding gas)</td></tr>
<tr><td colspan="3">Electro-slag welding (weld with filler wires and materials melted with the resistance heat of liquid slag)</td></tr>
<tr><td colspan="3">Electron beam welding (welding via the use of electron beams, a heat source with good convergent property)</td></tr>
<tr><td colspan="3">Laser beam weld (welding performed by using the energy of laser beams)</td></tr>
<tr><td rowspan="3">Pressure welding</td><td rowspan="2">Resistance welding</td><td>Lap resistance welding</td><td>Spot welding (welding by pressure bonding the joining contacting metal surfaces and then applying electric current to coalesce with the heat obtained from resistance to the current)
Projection welding (welding through heat obtained from resistance to electric current, using the shape of the contours of the contacting metal surfaces to be welded)
Seam welding (continuous welding by rotating roller electrodes and then applying pressure and electric current to continuously coalesce with the heat obtained from resistance to the current)</td></tr>
<tr><td>Butt resistance welding</td><td>Upset butt welding (welding by joining weld ends and then applying pressure and electric current to coalesce with heat obtained from resistance to the electric current)
Flash butt welding (making the surfaces of weld ends lightly come into contact with each other and then applying pressure and electric current to coalesce with heat obtained from resistance to the electric current)
Butt seam welding (welding by applying pressure and current to a portion of the butt surfaces of weld ends and then coalescing with the heat obtained from resistance to the electric current)</td></tr>
<tr><td>Fire welding</td><td colspan="2">Friction pressure welding (coalesce materials by using heat generated from solid body friction and then applying pressure)
Explosive welding (coalesce materials by making two kinds of metals collide at a high speed through the application of explosive force)</td></tr>
</table>

Fig. 1.1 Typical welding methods

Source: Author.

methods, as shown in Fig. 1.1, you will find various types, roughly classified into either of the two categories of fusion welding or pressure welding.

Fusion welding is a bonding technique that fuses components with the heat from shielding gas or a laser beam. Pressure welding, on the other hand, is a bonding technique that joins surfaces of matching components to each other (metals, etc.) by applying heat energy and pressure to fuse the atoms of the components.

The newly-developed welding technique is the seam welding technique made possible by pressure welding's resistance welding. In addition, the generally adopted method until now among automobile parts-makers has been spot welding by pressure welding. Thus, we will examine the features of the spot welding and seam welding methods, which are typical pressurizing techniques of resistance welding (welding by heating and pressurizing with the resistance heat generated *via* passing electric currents through welded joints).

Spot welding is a welding process that joins two components by layering and pressurizing them with a stick-shaped electrode and then passing electricity to heat the welded parts, causing them to fuse. This method is frequently applied in the welding of automobile bodywork, since it is applied for welding relatively thin plates (thin metal sheet, etc.). In addition, this method makes it possible to not only weld two metal sheet components, but more than three such components at a time.

Seam welding is a form of welding that pressurizes and passes electricity using roller electrodes and welding continually along the welded joint areas while spinning the electrodes. Since the area joined at one time in seam welding is minute, this method is suitable for ceramic or metallic package processing, whose welding operation requires only a low-level applied pressure and a small amount of electric current. For example, in crystal devices/Microelectromechanical systems (MEMS) and optical devices/Hybrid Integrated Circuits (IC), the seam welding technique is used to achieve a hermetic seal, covering ceramic and metallic packages with a lid (metallic).

Spot welding and seam welding are both resistance welding made possible through pressure welding, and the seam welding method can be considered to be the continuous application of the spot welding method.

Overcoming Challenges and Expanding the Usage of Seam Welding

The seam welding method has not been applied very much until now, despite its many merits. This is because the method has been criticized as follows:

1. Its electrical efficiency is poor, requiring a large electric installation; Due to its poor electrical efficiency, it cannot be applied for

general-purpose uses — it can only be applied for regular, fixed configurations;
2. Since it uses a large-capacity transformer, which is substantially large and heavy, welding operations become limited;
3. Awareness for the technology itself is low;
4. Since exterior water cooling is necessary and its service condition is bad, the method is not generally applicable;
5. The device itself is expensive and difficult to purchase.

At *Hikari Kogyo*, the engineers there have surmounted such seam welding challenges by developing a new method that passes electric currents through components by inserting them into ring-shaped electrodes and then joining them in the manner of sewing them together. This new method has increased the processing speed, thanks to the development of an original power source and power transformer for the welding machine, which assure the smooth stabilization and passage of heat-generating electricity.

For example, with regard to the components, just as with general welding machines, seam welding of stainless steel, galvanized steel sheets, aluminum, copper alloy, and titanium is possible. However, in the case of seam welding, the range thickness of such components it can handle has expanded from the limited conventional range of 0.5–2.3 mm to around 0.05–6 mm. In addition, with regard to the number of sheets it can handle, not only has it improved its handling capacity from two (differences in the thickness of boards of less than three times) to four layers of sheets, it has also boosted the processing speed range from the conventional 0.7–2 m/minute (thin plate 4 m/minute) to 0.7–10 m/minute (three layers of thin plates 10 m/minute).

In addition, the reduction in the mass of the welding apparatus to one-seventh what it used to be is also another major technical progress. In fact, the reduction in the weight of the equipment has also led to the adoption of seam welding in automotive factories, where it was rarely used before. Furthermore, by adopting seam welding in automotive factories, it has become possible to support high-level unit strengths even in thin steel plates, raising expectations for improvements in the reduction of automobile body weight and fuel efficiency.

In the automotive industry, since lightening of the automobile body is being pursed for the plug-in hybrid automobile and electric automobile, the applicability of the seam welding machine developed by *Hikari Kogyo* is expanding.

Bibliography

1. Nihon Keizai Shimbun, *Welding for Automobiles,* September 9, 2009.
2. *Art-Hikari* website http://www.art-hikari.co.jp/. (20120930)

2

The Reduction of Drying Time by 90%

Tetsuro Saisho

Applying the Technology of a Different Business

This section introduces a case that shows how a technology used in a particular industry helped a business in a different industry achieve a large-scale time reduction. The case shows the development and application of a drying machine used in the process of casting molds. Specifically, the machine uses microwaves, which project direct rays found in electrical radiation, to heat metals, such as iron, aluminum, and copper at temperatures higher than the fusing point to turn them into liquid, which are then poured into molds, cooled, and hardened into predetermined configurations.

Automobile parts-makers and industrial machinery-makers that use lost-wax casting need to congeal sand around solder models. Then, these models are soaked into a viscous liquid before sand is sprayed onto them. Every time this operation is carried out, there is a need to remove the moisture with a drying machine.

Lost wax casting is a sand-mold process and its mechanism involves creating molds by using a solder to make a model in the same shape as the finished product, solidifying its circumference with sand, and finally removing the solder. The conventional method of drying is by introducing warm air of about 25°C to dry a mold (of around 40 cm in length, 40 cm in width, and 20 cm in height). In total, this takes more than 50 hours.

Amid such a situation, the engineers at *Japan Highcomm Co., Ltd.* (Maebashi, Gunma, President Yukihiro Nishiyama), a company that produces devices that heat and sterilize foods by using microwaves, devised a mechanism that removes moisture by exposing molds inside a drying furnace to microwaves and warm air.[1] This was made possible by applying the company's microwave technology related to food processing.

The technology of microwave drying, which vaporizes moisture through microwave heating, has a wide range of industrial applications, ranging from food processing to construction materials and solid waste treatment. However, Highcomm has developed a technique that entails irradiating molds with microwaves and warm air in a drying furnace — a method of drying more effective than before.

With this method, the total time required for drying was cut down to three and a half hours, which is one-tenth of the time required previously. This is a reduction of one week in terms of delivery time. In addition, a space of at least several hundred square meters was necessary for hanging and drying the castings with the conventional method. However, since it has became possible to completely dry castings inside a furnace with the new method, the area required to carry out the drying process is reduced to just the area required for the installation of the apparatus, which is 2 m^2.

Various Drying Methods

Currently, there are various types of drying machines in use. There are many methods applied and the fundamentals behind each of them differ. Each also vary by the use of the drying machine. For example, in the case of the drying mechanism in dishwashers, the principle behind it is the use of hot, circulated air. The principle behind drying laundry in the sun, on the other hand, is the use of solar heat.

Water can easily be converted from a liquid to the gaseous state at a higher temperature. Its relationship with temperature is important and if the circumference of an article is dry, its moisture will evaporate more easily. Four other drying methods also rely on such properties of water.

The first is microwave heating, which dries by using the same heating apparatus found in general microwave ovens. Drying is carried out by the heat generated from vibrating water molecules caused by microwaves. This is similar to the internal heating of infrared rays. However in contrast to these rays, microwaves directly impact water molecules and thus drastically reduces drying time.

The second method is convection drying, which circulates hot air over the articles to be dried, eliminating moisture. In the case of hot-air drying, articles to be dried are arranged in a cabinet or on a shelf and drying is carried out in a stationary state. During the process, only blasts of hot air are used. Although this method require a lengthy drying time compared to suction-drying and infrared drying, it is simple in structure and can be reproduced at a low cost.

The third method is infrared drying which uses infrared rays for internal heating. In infrared drying, heated articles release steam, which increases the humidity as the steam fills up the space inside the drying machine. As a result, objects cease to be dried. The steam inside, therefore, has to be released through air blasts or suction to promote drying.

The fourth method is the suction-drying method, which is drying through leveraging the principle of maximum vapor-pressure. In suction-drying, water is evaporated at a low temperature by applying a low pressure. Drying is done by lowering the atmospheric pressure inside the drying machine and vaporizing the moisture content at low temperatures. When drying at low temperatures, the temperature of objects being dried is not high, so it is possible to dry without having the heat degrade weak materials, avoiding any damage to the objects being dried.

A Device Designed to Meet Clients' Demands

Since more than 10 years ago, *Japan Highcomm* has been engaging in the proprietary development and commercialization of microwave-based heat sterilization and thawing equipment for foods. The company made its entry into the field of microwave devices, where preexisting devices had been commonly used.

Japan Highcomm has leveraged its production technology to offer microwave equipment that meets the demands of its clients (see Fig. 2.1). Namely, the company's production technology has been driven by its engineering business that mainly focuses on product line engineering, industrial plant design and manufacturing, production technology consulting, and microwave facility design and production.

Fig. 2.1 A drying furnace using microwaves

Source: *Japan Highcomm's* website.[3]

The opportunity for developing such a device came about when a tofu maker entrusted the company with the task of developing a sterilizing apparatus that uses microwaves for its new products. Since, the device saw its application expand from just heat sterilization and thawing to drying and cooking. In 2010, the company also entered into the field of industrial applications for the automobile industry, when it introduced the microwave-based casting mold dryer.

In the microwave apparatus generally used in microwave ovens at homes, a cathode is placed at a cylindrical central axis and a strong magnetic field is added axially to generate oscillations of microwaves, and using a diode magnetron, the water molecules on the foods are vibrated by heat. A feature of the thawing unit that uses heat to kill bacteria is its capability to optimize microwave-irradiation settings to levels that most suitably meet the conditions of food size and weight. In other words, since it is possible during thawing to decrease the outflow of moisture or amino acids in foods, defrosting can take place without compromising flavor. In addition, with this device, even the time required for thawing, which involved several hours previously, is reduced to between several to 10 minutes, which enable shorter delivery times while making it possible to easily adjust production quantity.

The emergence of such a microwave equipment is made possible by the crux of the company — its engineering business. *Japan Highcomm's* technology designed for food processing became popular, and by assuming the leading role in lines related to automotive factories and other manufacturing concerns that make use of this technology, the company was able to develop the microwave-based drying machine, drawing on its expertise and experience (software) and its technology (hardware). Essentially, the company developed the mold drying microwave-based machine in collaboration with casting firms and lost-wax casting makers, applying its microwave technology designed for processing foods to the process of drying molds and innovating a machine that applies microwaves and warm air to molds placed inside a drying furnace.

Lost-wax casting is a form of molding that duplicates the shape of a product with a solder, covers the duplicate's circumference with sand, and pours metal into its hollow in which the solder has been eliminated by dissolving it. In most cases, this form of mold drying takes a long time with temperature and humidity being managed with a thermostatic chamber. The company was able to revolutionize conventional thinking behind drying by using a microwave-based machine for mold drying.

Although there have been attempts made in the past to use microwaves for drying purposes, analysis of pre-existing products using the microwave technology showed that such attempts had failed. Japan Highcomm, on the other hand, was able to develop the microwave-based mold drying machine because the company had accumulated technologies, expertise and experience in various food-related fields, and was then able to reapply them.

Bibliography

1. Nihon Keizai Shimbun, *Microwave-Based Mold Drying Machine,* October 20, 2010.
2. Organization for Small & Medium Enterprises and Regional Innovation, JAPAN (SME Support, JAPAN) "J-Net21, *Deployment of Microwave Equipment for Industrial Use in the Food Sector* (*Japan Highcomm*) May 17, 2011 Retrieved from http://j-net21.smrj.go.jp/well/genki/2011/05/post_491.html. Accessed on September 30, 2012.
3. *Japan Highcomm* website. http://www.highcomm.co.jp/index00.html (20120930).

3

The Sudden Rise in Web Conferences and Teleconferences

Akira Ishikawa

Japanese Market Size: 70 Billion Yen

Web and video conferences between and within companies, as a form of visual communication, are rapidly increasing. By 2007, the market in the United States had expanded to 120 billion yen and the 2008 survey carried out by Seed Planning Corporation (Taito-ku, Tokyo, President Yoshio Umeda) predicted that the Japan market worth 70 billion yen would have grown exponentially to exceed 700 billion yen in 2010.

So why are we seeing such a rapid rise in the use of web conferences and teleconferences?

The primary reasons are firstly a reduction of transportation expenses and secondly the elimination of traveling times. But that's not all. As companies become increasingly eco-friendly, they become aware that these means of communication help to cut down on CO_2 emissions and provide a way to prevent environmental deterioration. What's more, in the event of a pandemic, these means of communication can also serve as effective measures.

The question is whether, by using this kind of computer network-assisted conferencing system, a company can achieve its original business objectives without seeing any drop in operating effectiveness, and whether doing so will lead to sufficient profitability.

With regard to the question of improving operational efficiency and productivity, *V-CUBE*, Inc. (Meguro, Tokyo, President Naoaki Mashita), a company selling "Nice-to-Meet-You Meeting", a web-based conferencing system, reported that the adoption of these means of communication have led to an improvement in sales because they have helped to revitalize office communications, pointing out how much easier it has become to convey details to clients at remote distances — details that were otherwise difficult to convey over the telephone or *via* email.

Such details include more than just language, but also nonverbal messages, such as their manner of speaking, actions, gestures, the pauses in communication, and context.

Research conducted by experts of non-verbal communication have found that a message conveyed *via* verbal language in the course of a two-person interaction amounts to only 35% of the overall message, and the remaining 65% is made up of forms other than language, as stated above.

A Three- to Six-Fold Improvement in Level of Understanding

Furthermore, according to a survey carried out by the U.S. Air Force, compared to communicating solely with just language, pictures or other reference materials, a person can improve three- to six-fold in comprehension and memory retention when pictures and other reference materials are used in tandem with language.

Next, in terms of profitability, according to *V-CUBE's* seminar materials, the quarterly results of Company W showed that the company saw a dramatic decrease in transportation time when they adopted *V-CUBE's* system, registering a total of 34 hours in contrast to 162 hours when they were not using the system. Furthermore, upon closer examination, *V-CUBE's* data also disclosed that the time required for branch-manager meetings went down dramatically from 102 hours to just 34 hours, and the time required for making a visit to the managing director went down from 60 hours to zero hours. Their expenses went down, although not as dramatically or exponentially as the changes mentioned above, from 926,280 yen to 480,800 yen, achieving a total cost saving of 445,480 yen.

As indicated in Fig. 3.1, this company revealed that the means of visual communication made possible *via* the web-based conferencing system is a key tool that fills the lulls in communications conducted over the

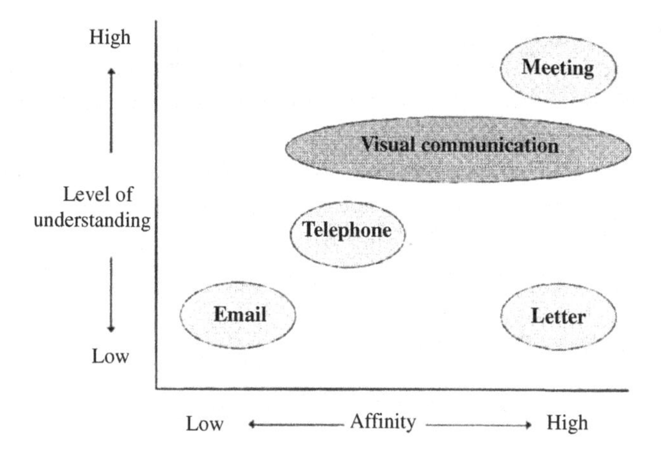

Fig. 3.1 Position of visual communication

Source: Seminar material from V-CUBE

telephone, email, and even letters. In fact, the company asserted this tool to be key, even though there is a decrease in psychological distance and level of understanding compared to face-to-face interactions.

Incidentally, the conferencing system supported by computer networks can be categorized as either supported by a "special — Purposed" Machine or software-based. The former is a setup for what is generally called teleconferencing while the latter is the web conferencing system we are discussing here.

Since teleconferencing uses specialized telecommunication hardware, its image is extremely sharp and TV-like in quality. On the other hand, however, the cost for installing a unit is relatively high, and since users need to go to where the unit is installed, it is not possible to make use of it anytime, anywhere.

Drastic Cost-Cutting

On the one hand, web-based conferencing is inexpensive and can be used on a monthly subscription basis (*V-CUBE* charges a basic monthly fee of 79,900 yen under its standard plan) with existing services, allowing anyone to use it with an internet connection, even from one's office desktop or from business-trip destinations. On the other hand, however, since the conferencing system makes use of PCs used in everyday business operations, depending on the maker, software installations may be necessary, which require companies to establish an internal company network.

Nevertheless, in the event of having a face-to-face business talk, companies need to consider not only travel costs or personnel expenses associated with the travel, but many other risks such as unanticipated delays or accidents that could occur in the course travelling, and even the likelihood of one party becoming absent due to illness or the need to attend to an urgent matter.

4

The Reduction of Traveling Time from Five to Three Hours

Kazuo Matsude

Launch of the Kyushu Shinkansen (Bullet Train)

The Kyushu Shinkansen service (between Hakata and Shin-Yatsushiro), operated by the *Kyushu Railway Company* (*Kyushu Railway Co., Ltd*, Hakata-ku, Fukuoka-shi, President Koji Karaike), started on March 12, 2011, when all lines of the Kagoshima route started as well. Unfortunately, as Japan was hit by the Tohoku earthquake and tsunami disaster — a severe disaster of an unprecedented scale — a day before the inauguration, the celebratory mood building up to that day dissipated and the launch was carried out under a subdued atmosphere.

The Kyushu Shinkansen links Kagoshima Central Station with Hakata Station, from which it connects to the Sanyo Shinkansen bound for Shin-Osaka. The Shinkansen falls into two types; one is the 800 series (running within the Kyushu area, having a maximum speed of 260 km/hour) and the N700 series (Sanyo Shinkansen train, having a maximum speed of 300 km/hour). The Kyushu Shinkansen's maximum speed is slightly lower because slopes and curves were not dealt with when constructing its rail tracks to reduce construction costs.

Even then, compared to conventional lines, the effect this new line has on shortening travel times is remarkable. According to the timetable of March 1989, the time required to cover the distance from Nishikagoshima (currently, Kagoshima Chuo) to Hakata used to be 246 minutes, using the Super Ariake line, which was the fastest line available at the time.

Subsequently, with the partial start in March 13, 2004 of the Shinkansen service between Kagoshima Chuo and Shin-Yatsushiro, the time distance was shortened to 132 minutes. However, with the Mizuho, a newly-launched line, this same route is covered in a matter of only 79 minutes, realizing, effectively, a super effect/productivity surge.

With this result, the psychological sense of distance between Kyushu and Kansai has also shrunk substantially. For example, the time required to cover the distance between Kumamoto and Shin-Osaka until now had been 237 minutes at minimum. However, with the new line, it now takes only 179 minutes, reducing traveling time to less than three hours. Similarly, the time required to cover the route between Kagoshima Chuo and Shin Osaka has been shortened from 302 minutes to 225 minutes, with traveling time reduced from five to three hours.

In the transport industry, insiders commonly mention a "four-hour wall" when talking about moving from point to point. This refers to the fact that many people would only choose railway travel over airway travel when the transit time required is less than four hours.

According to this line of thinking, with the launch of the Kyushu Shinkansen, the passenger traffic between the Kyushu-Osaka route will see a considerable number of aircraft users switching to traveling *via* railroad.

In addition, although it still takes more than four hours to travel from Kagoshima to Kyoto or Nagoya, judging by the fact that traveling within Kyushu has become markedly faster and by the fact that the Kyushu Shinkansen is drawing public attention, it is anticipated that the demand for package tours of the following kind will rise: tours that feature a one way trip by bullet train to Hakata from these cities (taking less than four hours), access to tourist destinations within Kyushu *via* the Kyushu Shinkansen, and a return flight from Kumamoto or Kagoshima (or this route in reverse).

Cross-Border Marketing

In this way, the bullet train is very likely to become the main means of transportation for travel within the area west of Osaka. However, since networks of conventional lines and expressways are developed to a considerable degree within the domestic transportation network, the launch of the bullet train can be said to have merely raised the level of the convenience of these networks instead of having attained a predominant position.

Thus, the Kyushu Railway Company, hoping to further extend its market, is putting a great deal of effort into reinforcing accessibility to the rest of Asia, or Korea in particular. When you examine a map, you can reaffirm that Kyushu and Korea are in close proximity to each other. The distance between Hakata and Busan is approximately 200 km. The Kyushu Railway Company, in cooperation with Miraejet in Busan, operates a hydrofoil, high-speed ferry, offering access to Busan with a time distance of 175 minutes. In the period from 2004 through 2008, the number of passengers varied annually between 500,000 to 600,000, among which 250,000 through 300,000 were Koreans, accounting for almost half the total.

The ship only ferry travelers and since Japan and Korea do not officially permit each other's vehicles to enter each other's ports, "the roadways, in effect, remain unconnected." Consequently, there is no immediate competition with any network of expressways, unlike within Japan, making this situation advantageous from a marketing perspective. Furthermore, there is the Korea Train Express (KTX), the rail system that services the Busan-Seoul route, covering a distance of 423.8 km in 138 minutes at minimum.

The technology behind KTX, unlike Japan's Shinkansen, is based on the power-car traction technology of France's rail service, Train à Grande Vitesse (TGV). Launched in April 2004, the KTX train is large in scale, comprised of twenty cars; two traction heads, or powered ends, and eighteen passenger cars, boasting a maximum speed of 300 km/hour. It started its Seoul to Busan service in November 2010, linking the two areas *via* Gyeongju, the most popular tourist destination in the nation.

When comparing the urban population in Korea, linked through high-speed ferries and the KTX, with the urban population of cities found along the Sanyo Shinkansen's railway line, one could see the vast potential market (Fig. 4.1).

As indicated by the table above, even though KTX's total distance amounts to only two-thirds of Sanyo Shinkansen's total distance, KTX's total population of major cities found along its railway route is approximately three times more than Sanyo Shinkansen's total population of major cities found along its route. For the Kyushu side, the expectations are high for tourism marketing that targets Korean tourists.

And it may be said that the Kyushu Shinkansen is assuming the role of "a cheerleader" for promoting such a cross-border marketing initiative. For this purpose, the Kyushu Railway Company's website is also available in English and Korean.

Along KTX's railway line (Total distance 423.8 km)		Along Sanyo Shinkansen's railway line (Total distance 644.0 km)	
Busan	3,498	Osaka	2,629
Daegu	2,457	Kobe	1,525
Daejeon	1,496	Okayama	675
Seoul	10,032	Hiroshima	1,154
Total	17,483	Total	5,983

Fig. 4.1 A comparison between populations of cities found along KTX's railway line and populations of major cities found along Sanyo Shinkansen's railway line (Unit: 1,000)

Source: Based on data obtained from the United Nations'.[1]

Concrete Tourism Promotion

Then what specifically is a tourism promotion? To cater for tourists from Seoul and Busan who travel to Kumamoto and Kagoshima, it is difficult to offer up travel options that take less than four hours by land and sea routes. Alternatively, if a Korean tourist were to use a Shinkansen bullet train line, first he/she must go to Hakata (Fukuoka) as his/her Japan-side access point, while including it as one of the sight-seeing destinations.

By doing so, it will become possible to reduce the time required for travel, including the time required to transfer from a high-speed ferry to a Shinkansen bullet train. Furthermore, by incorporating the Kyushu Shinkansen itself as a form of tourism attraction, it is expected that the one-way use of an aircraft as a means of transportation for exit (or entry) will become a natural choice.

In this case example, we saw how the super effect/productivity surge was achieved by combining existing technologies related to high-speed railways while keeping profitability in mind. Subsequently, the super effect ended up being associated with Kyushu's regional uniqueness — a factor that spurred the railroad business to explore overseas opportunities. In this respect, it can be viewed that the pursuit for the super effect/productivity surge has led to a new venture.

Bibliography

1. United Nations, *Demographic Yearbook,* United Nations 2008. http://unstats.un.org/unsd/demographic/products/dyb/dyb2008.htm. Accessed on April 9, 2011 (20110409).

2. Kyushu Railway Company http://www.jrkyushu.co.jp/korean/time_table/time_table.html (20110407).
3. Toshihiko Aoyagi, *Launch of All Kyushu Shinkansen Lines,* Gakushikaiho, Gakushikai No. 887, 2011.
4. Susumu Ishihara, *Launch of all Kyushu Shinkansen's Kagoshima Route,* Gakushikaiho, Gakushikai, No. 887, 2011.
5. Ryozo Kawashima, *Full-Scale Investigation into The Nationwide State of Railways — The Kyushu Edition (1),* Soshisha Publishing, 2006.
6. Konest "KTX," *Korean Transportation,* Konest, 2010. http://www.konest.com/data/traffic_info_detail.html?no=1230 (20110406).
7. Yoshitaka Mizoo, *Tourism Studies — Fundamentals and Practice.* Kokon, 2003.

5

Shortening Inspection Time by 90%

Akira Ishikawa

Definition of Radio Frequency Identification (RFID)

In this section, I will discuss the technology of Radio Frequency Identification (RFID), which is a way to identify objects using electromagnetic waves. In most cases, it refers to technologies that can acquire/transmit information obtained from tags containing ID information, *via* short-distance communications using electromagnetic fields or radio waves.

Specifically, a winning combination is to have Radio Frequency (RF) and Integrated Circuit (IC) tags converge with the wireless technology of readers that can read its data. However, since such tags can be attached to various objects and even onto human beings, they can be useful in monitoring and verifying positions and shifts in positions, and therefore reducing inspection times.

Speedy and accurate distribution support

An example is the distribution reforms through the application of a monitoring and tracking system for pharmaceutical products *via* the use of RFIDs by *Eisai Distribution Co., Ltd.* (Atsughi-shi, Kanagawa, President Jiro Kimura) and Sato Holding Corporation (Meguro-ku, Tokyo, Chairman Koichi Nishida) won the 11th Automatic Recognition System Grand Prize, which gives public recognition to advanced systems and technologies that demonstrate remarkable effects. This system works by monitoring and tracking *via* RFID technology highly accurate and efficient inspections of deliveries and arrivals at pharmaceutical factories, distribution centers, and agency delivery centers. It also enables the swift and correct responses to inquiries concerning the transportation and delivery of shipments after they have been dispatched.

A feature of the system include a one-way use in the original packing (which is easy to install) of psychotropic drugs, which require strict control. It is also designed for shared use among several companies; in this case, factories, distribution centers, and agencies.

Zero Error

As part of a "product substitution experiment" (which was carried out without informing company workers), surprise inspections, which involved gauging the accuracy of human inspections were carried out for a period of two months. The results registered a 4% error rate. In contrast, when RFIDs were used, the results showed a completely error-free, 100% accuracy rate.

In terms of efficiency and time saved, the IC tags succeeded in producing a super effect/productivity surge, as indicated in Fig. 5.1, which shows a comparison of inspection times. What had taken 20 to 30 minutes for visual

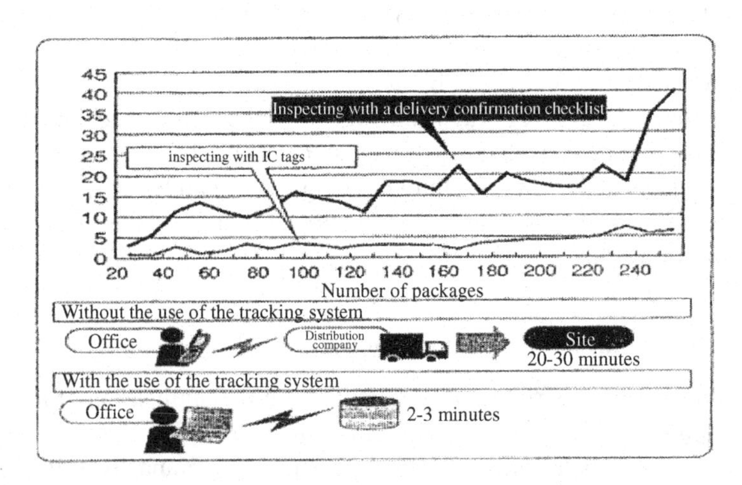

Fig. 5.1 A comparison of Inspection-time (x axis — Number of packages; y axis — Inspection times)

Source: Japan Automatic Identification System Association, "Findings of Trends in the Automatic Identification Apparatus Market — Although down by 9% in 2007, a slight increase is forecast for 2009", "JAISA Bulletin," 2009 Summer Edition, Vol. 11/No. 2, 2009.

inspection (with the aid of a delivery confirmation checklist) was reduced to only 2 to 3 minutes when IC tags were used to carry out the inspections.

But there was more to the super effect. It raised the sense of security, making it less stressful for addressees and delivery workers. The reduction of stress for patients, their families, and for people working in medical institutions, particularly licensed nurses, is also priceless.

According to the 2008 report compiled by the non-profit organization, Japan Council for Quality Health Care (Chiyoda-ku, Tokyo, Chief Director, Tetsuo Ihara), there were 1,440 medical *accidents* reported and a total of 220,000 medical *incidents* reported as well. While medical *incidents* do not amount to medical *accidents*, it is still extremely desirable to see the number of such incidents decrease even if it is just by a little.

All the more, when an error occurs, time is also lost to remedy the error, so this is also a factor that cannot be ignored.

The Pros and Cons of Two Methods

When classifying RFID tags by power source, we can classify them into the passive or active type. The former does not have a battery inside and operates by adjusting electromagnetic waves emitted from reader-writers, so they are cheap and do not require much maintenance. But their drawback lies in their short communication range, extending only several meters.

The latter, on the other hand, operates by drawing power from the battery in it, so its communication range extends much further — no doubt a super effect. However, its drawback is that it is limited by its battery life, thus requires maintenance. Furthermore, it is relatively expensive.

When examining the RFID tag by frequency band, another super effect becomes evident here as well. In the case of the long wave frequency band (below 135 KHz), the maximum possible communication range extends to 0.3 m. Although a relatively long antenna is required, it is resilient to the impact of water and dust.

In contrast, in the case of the microwave frequency band (2.45 GHz), the transmission method changes from the electromagnetic induction method to the Hertzian ray method and the communication range extends to 2 m, a one-digit difference.

Although a short-length antenna suffices, this has the drawback of being vulnerable to the impact of water and dust.

Since the RFID tag's performance markedly varies by the power source it uses and by its frequency band, it becomes necessary to make use of it in a way that leverages its characteristic properties. However, compared to the conventional bar code, the RFID is clearly the winner in many ways; in terms of recordable data capacity, which is larger by two digits, amounting to several kilobytes; in terms of its maximum communication range, which is comparably larger (single-digit meters vs. dozens of centimeters); in terms of its ability to make simultaneous readings of multiple sources; and in terms of its penetrability (identifying individual units behind obstacles such as cardboards).

6

Instantaneous and Spontaneous Information Sharing

Hiromichi Yasuoka

The Impact of Social Media at Times of Emergencies

On March 11, 2011, the Tohoku Earthquake struck Japan. During the state of emergency, social media such as *Twitter* and *Facebook* played active roles as sources of information. Users in disaster-stricken areas and in various regions of eastern Japan, and even in other areas, used their cellphones to report, just like journalists, a continuous moment-by-moment account of their situation. In addition, in response to tweets from users who requested the widespread dissemination of their information, celebrities re-tweeted those tweets and helped to inform a large number of people who were following them (individuals on their friend list). In other words, a celebrity had become a type of mass media in his or her own right.

What such social media proved, during this earthquake disaster, was that instantaneous and spontaneous sharing of information was possible. Indeed, they proved to be an effective tool for disaster control, realizing a super effect/productivity surge in information sharing. Firstly, private enterprises were using them, and during the aftermath of the earthquake disaster, there were companies actually using *Twitter* and *Facebook* to track updates on safety information.

Private enterprises leverage the efficiency of social media's information transmission and collection capabilities to carry out; promotions and branding, public relations, product planning, product sales, product improvements, CRM/customer support, and talent acquisition (Human Resources).

Use of Social Media by Local and Central Government Offices

Even local and central government offices can utilize social media, which are essentially tools for real-time transmissions of information. While a static type of media such as the website is probably more suitable for making regular announcements, for the purpose of transmitting updates on a constant, real-time basis, social media proves more suitable.

During the initial period of the earthquake disaster, the disaster-prevention radio of Kamaishi city streamed the forecast that the height of the tsunami would be 3m. This information ended up spreading on its own. During such an emergency, there is a need to update information on a timely, moment-by-moment basis, and because there is also a need to make the public aware of the situation as soon as possible, it is necessary to keep streaming it even if it happens to be ambiguous to some degree. When a media continuously streams in this way, users will very likely attempt to pursue the information being disseminated. Therefore, if the initial information is deemed to be lacking (or exaggerated), it is easy to make revisions to it. Since such urgent cases can occur unexpectedly in the future again, utmost awareness is needed so as to serve as more general points of reference.

Furthermore, even local and central government offices can make use of social media as tools to collect information on a real-time basis. It is necessary to closely examine information by capturing content from various places transmitted by various users, and to select what is relevant by comparing them to your own information. A number of information-collecting schemes already adopt such practices. One such example is the weather report, which produces weather forecasts by drawing on various regional reports from weather reporters located all across the nation. Application of social media as a tool for gathering information should be extended for not only the weather forecast, but also early-stage conditions of disaster-stricken areas. As users become reporters, a super effect in the deployment of reporters can be seen.

As seen in the case of private enterprises, social media can be helpful in collecting information for the purpose of verifying the personal safety of users. If an individual is seen to have sent out information, the safety of

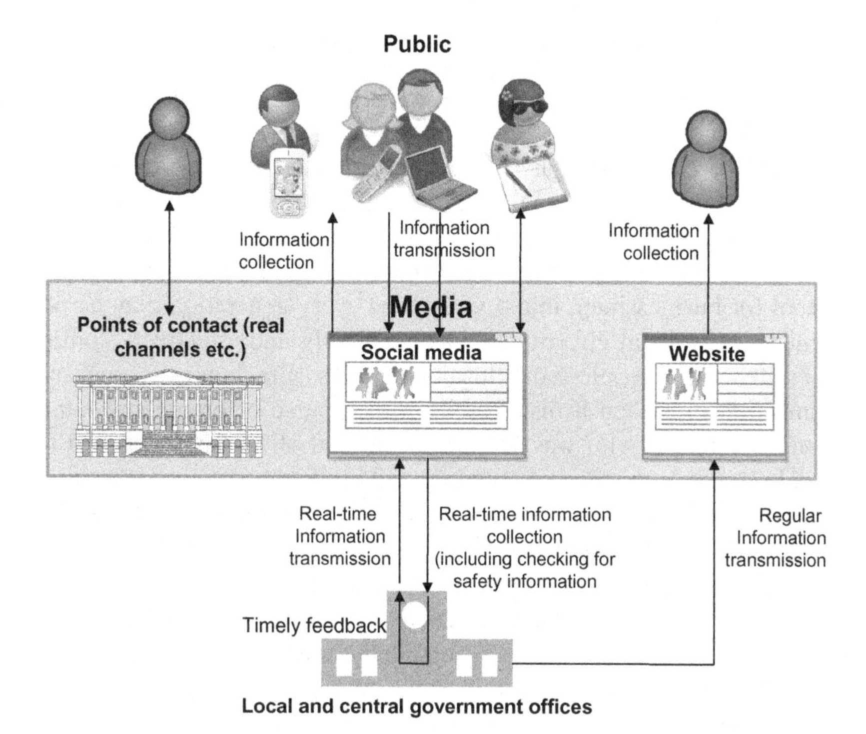

Fig. 6.1 Cycle of ideal information collection and transmission
Source: Author.

this individual is practically confirmed, eliminating the need to send out an email to him or her to confirm his safety and waiting for a reply. It is also possible to verify the safety of registered staff, other authorized personnel and citizens by collecting positional information.

Central and local governmental bodies could filter these pieces of collected information and link the relevant ones to future information dispatches. Naturally, the reliability of some information may be called into question, but users (readers), who are anxious that their locales may become be struck by disaster, desire to grasp information as early as possible. Therefore, while those information that are understood to be disinformation or rumors (widespread baseless information) should be eliminated immediately, useful information that should be obtained even though they cannot be confirmed, should be released as "speculation". As shown in Fig. 6.1, how fast one can complete such a cycle will become key in realizing effective use.

The Use of Social Media is a Demand of the Times

Various confirmations of the daily existence of an individual can be made by examining his or her usage history of social media, along with his or her usage of electricity, gas, water, and other records of activities left behind as digital data (life logs). This applies to people below a certain age, since social media has yet to catch on among the elderly.

However, since it has become indispensable as an information gathering tool for businessmen, in ten years time every generation may be proficient users. A local government can constantly and effectively confirm the existence of people from log-in histories (life logs), as one such organization did when it discovered, in the course of organizing family registers, a person who was 150 years old and alive. They checked the log-in histories of this person (life log) and verified his status (after ruling out fraud). This could be a good way to restore credibility in the information from the government.

In addition, the local government of Ibaraki, which has one of the largest *Twitter* followers among local governmental bodies in Japan, held a contest *via YouTube* to improve its regional brand image. They were able to double their visitor-traffic number by running a campaign that linked their website to their *Twitter* and blog posts and effectively improve their brand image.[1,2]

Furthermore, due to the impact of the Tohoku Earthquake and Tsunami Disaster, the number of local governmental bodies using *Twitter* as a means of carrying out PR activities increased. Specifically, as indicated in Fig. 6.2, the number rose from 121 (as of March 11, 2011) to 148 (as of April 4 of the same year).[3]

There are plenty of such users of social media. For example, there are already around 20 million users on *Twitter* and the domestic social networks, *mixi* and *GREE*, and *Facebook* has nearly 10 million. Simply adding the users of these social media yields a total of 70 million, which is more than half of the population of Japan.

Of course, if you take into account overlapping accounts and inactive users, the actual total may be half or one-third of this amount, but the number of users is seeing an upward trend, and it is not premature for local and central governmental bodies and business entities to consider carrying out their own social-media measures.

In addition, there has been an appearance of a service called *paper.li* (www.paper.li), which publishes not only for business organizations but individuals as well, daily summaries in newspaper format of information

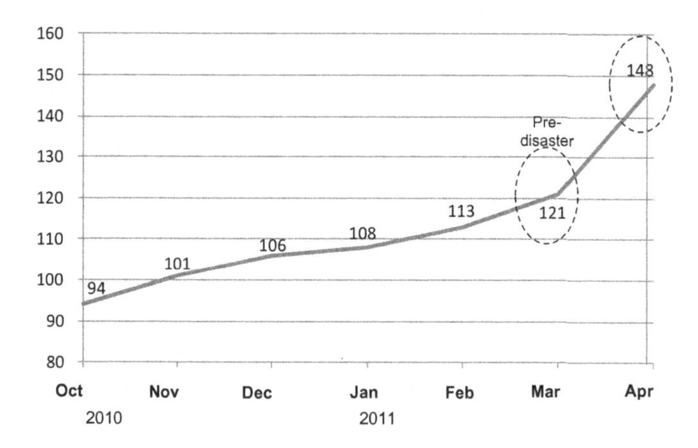

Fig. 6.2 Fluctuations in the number of the *Twitter* accounts of central and local governmental bodies

Source: Ministry of Economy, Trade and Industry.

trending through *Twitter* and *Facebook*. It is not an exaggeration to say that, thanks to this service, newspaper companies and publishers exist in more parts of the world than ever.

Furthermore, *Facebook* has helped trigger revolutions. It is well-known that it has played a role in initiating the revolutions that occurred in African countries such as Egypt. Their inceptive role was sufficient to produce a super effect in the realm of information sharing.

Bibliography

1. Seed Planning, *2011 social media and the latest trends in the regional vitalization business secto — points on social media usage, learned from precedents,* Market research report, Seed Planning, April 14, 2011.
2. Tomohisa Ito/Hiromichi Yasuoka/Katsumi Tomita/Natsuhiko Sakimura, *Private-enterprise platform strategy for advancement of open governance — Realization of open central and local governments initiated through social CRM, Intellectual property creation,* Nomura Research Institute, July 2011 edition, 2011, pp. 4–19.
3. Ministry of Economy, Trade and Industry, *The creation of the Twitter account recognition scheme for public institutions,* April 14, 2011. http://www.meti.go.jp/press/2011/04/20110405004/20110405004.pdf. Accessed on May 30, 2011 (20110530).

7

The Development of a Three-Minute Battery Charger

Toru Fujii

With the priority given to building an environmentally sustainable society, the introduction of the plug-in hybrid automobile and the Electric Vehicle (EV) is drawing worldwide attention. They both have superior performance in terms of energy efficiency and cutting down on CO_2 emissions.

The mainstream charging system for the EV in the future will make use of an external power source, making the installation of charging facilities a major issue.

Even in Japan, according to the energy master plan (as of the cabinet decision made on June 18, 2010), the number of charging facilities targeted for installation by 2020 is two million for standard charging facilities and 5,000 for express charging facilities.

Power receiving systems are available in two kinds: the standard vs. the express power receiving system. For the technology behind express charging, there is the *CHADeMO* system, which is an externally installed express charging system adopted by Japan, and the AC system, which is developed in Europe and uses the motor-controlled inverter. The Japanese system is one step ahead of the European one. A summary of the features of the charging systems appears in Fig. 7.1.

As shown in Fig. 7.1, the difference between a standard charging system and an express charging system is drastic, with the standard charging system taking approximately seven to fourteen minutes while the express charging system takes approximately thirty minutes. This is approximately 10 to 20 times faster, no doubt a super effect. Additionally, the difference in price of

Types of recharging systems		Normal charge			Boosting charge
		Electric outlet		Pole type standard charger	
		100V	200V	200V	
Assumed charging locations (examples)	Private	Detached houses/apartments, buildings, outdoor parking spaces, etc.		Apartments, buildings, outdoor parking spaces	- (Very restrictive)
	Public	Car dealers, convenience stores, hospitals, commercial buildings, pay-by-the-hour parking lots			Roadside stations, petrol stations, SA freeways, car dealers, commercial buildings
Recharge times	Cruising range 160km	Approximately 14 hours	Approximately 7 hours		Approximately 30 minutes
	Cruising range 80km	Approximately 8 hours	Approximately 4 hours		Approximately 15 minutes
Base price examples of charging facilities (construction expenses not included)		Several thousand yen		Hundreds of thousands of yen	1 million yen and above

Fig. 7.1 Types and features of charging systems

Source: Ministry of Economy, Trade and Industry data/Ministry of Land, Infrastructure and Transport data.[2]

the battery chargers is also drastic, with the battery for standard charging costing from several thousand to several hundreds of thousands yen, whereas the cost of a battery for express charging is more than one million yen.

The essential issues that need to be addressed for the practical application of charging systems in the future is to achieve further reductions in charging times and realizing lower prices, not to mention improving operability and safety.

The Successful Development of Ultra-Fast, Three-Minute Express Charging

Amid such technological trends, the JFE Engineering Corporation (Chiyoda-ku, Tokyo, President and CEO, Sumiyuki Kishimoto) announced that they had succeeded in developing an "ultrafast battery charger" that can recharge an EV in a matter of three minutes.

The company premised their development concept on the idea that EV users would desire charging to take three minutes, which is around the time it takes to refuel at a gasoline station or to finish shopping at convenience stores. Consequently, they set the target of three minutes and succeeded in developing an ultra-fast charger. As a result, they were able to invent a technology that could charge an EV's battery up to 50% capacity in three minutes, and up to 70% capacity in five minutes.

According to JFE Engineering, the "50% charge" means that the EV will be able to run up to 80 km, which is half of the 160 km it can run on a full charge. Also, in five minutes, a 70% charge can be achieved, enabling the EV to cover the distance of approximately 110 km. This distance is considered to be longer than the distance a privately owned car usually covers in a day on average, and when you take into account of the fact that conventional express chargers can only charge up to approximately 20% in five minutes, you can see that the development is epoch-making, bringing about a super effect in time reduction.

As indicated in Fig. 7.2, the ultra-fast charger incorporates a built-in storage cell inside the battery charger that stores, over a period of several

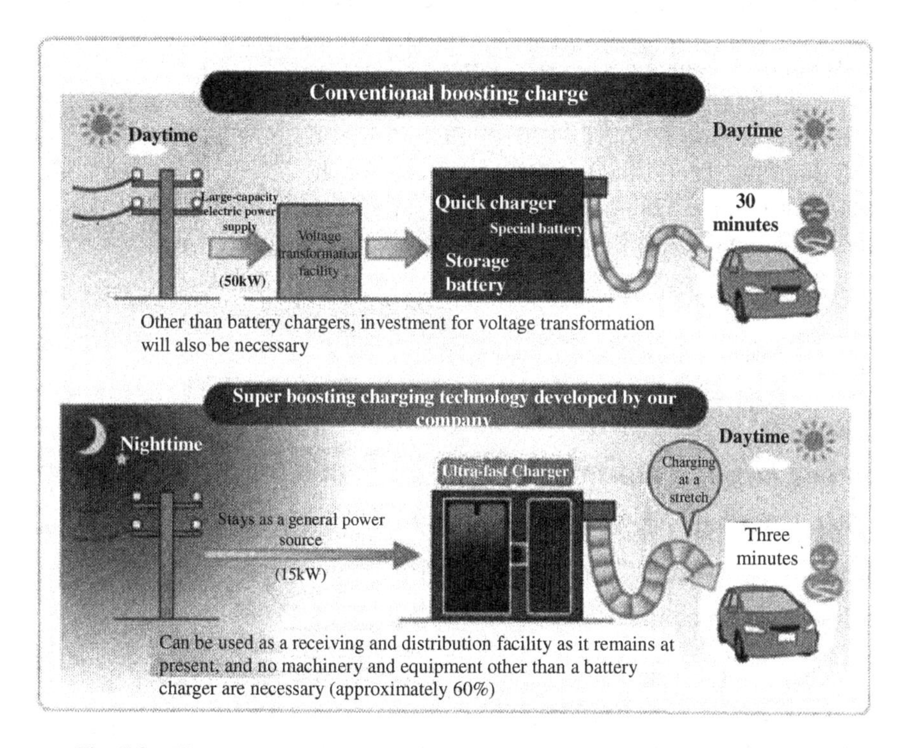

Fig. 7.2 The mechanism behind JFE Engineering/Super-Express Charger setup
Source: Nikkei BP Eco Management Forum.[3]

hours, electricity generated in the night. Thereafter, this stored electric power is drawn up by another special battery to instantly charge an EV. This special battery has made it possible to charge in a super-short period of time.

At present, to install a battery charger at a service station, a capital investment of nearly 10 million yen is necessary to pay for not only the main unit of the express charger, but also for the facilities that reinforce power-receiving capacity.

Conversely, the super-express charger does not require reinforcement facilities and since its capital spending cost can be brought down to approximately 60%, the initial cost is reduced and the running cost is also reduced because the electricity it uses is generated at night, when electricity rates are economical. Furthermore, the super-express charger has many other merits, such as its potential to level out electricity consumption.

JFE Engineering has launch in June 1, 2010 the "Super-Express Charger Project Team" to advance the commercialization and industrialization of its product and carry out demonstrations and experiments for its early market launch.

In addition, with regard to the spread of the EV in the future, judging that the convenience of the battery charger will become key for the user, and that the issue of cost will become key for the installer, the company has developed a technology that drastically reduces the charge time, as shown in Fig. 7.3, while reducing the cost for the required facilities.

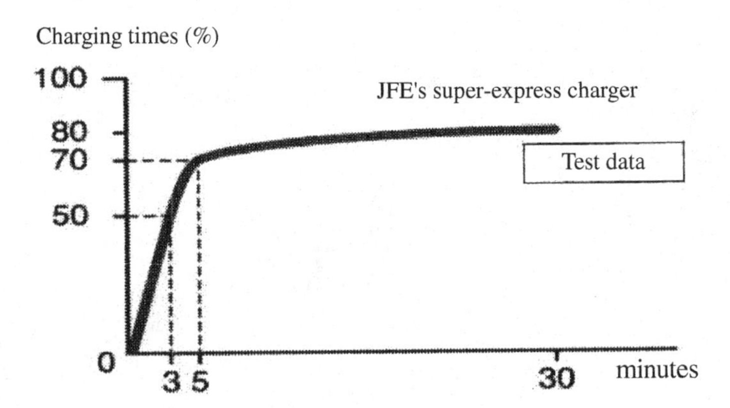

Fig. 7.3 JFE Engineering/Relationship between charging times and charging rates

Source: JFE Engineering.[1]

From now on, spearheaded by the Super-Express Charger Project Team, the company's initiative is slated to get into full swing. Furthermore, the company also aims to have installations at gas stations and convenience stores.

Bibliography

1. JFE Engineering news release, *Establishment of an organization to advance the industrialization of a super-express charger for electric automobiles — A practical three-minute charging system will pave the way for the future of the electric vehicle*, JFE Engineering, June 1, 2010. http://www.jfe-eng.co.jp/release/news10/news_e10006.html. (20110516)
2. Ministry of Economy, Trade and Industry/Ministry of Land, Infrastructure and Transport, *Guidebook for installing charging facilities for electric automobiles/plugin hybrid automobiles*, Ministry of Economy, Trade and Industry/Ministry of Land, Infrastructure and Transport, December 7, 2010. http://www.mlit.go.jp/common/000130718.pdf. (20110516)
3. Nikkei BP Eco Management Forum, *JFE Engineering's express charger charges an EV in 3 minutes*, Archive article, ECO JAPAN, June 3, 2010. http://eco.nikkeibp.co.jp/article/news/20100602/103948/. Accessed on May 16, 2011. (20110516)

8

High-Speed Measurements Enabled by the Multi-Head Weigher (The Computer Scale)

Takashi Yonezawa

Manual Measurement

How can you efficiently process vegetables and candies into packages of uniform weight? Taking up this challenge, a company has invented a proprietary technology in terms of packaging speed and accuracy. When packing liquid or powdered foods into bags, it would be appropriate to just keep filling until the target amount is reached, making fine adjustments to the quantity at the end of the process. However, when packing food items that vary in shapes and sizes, it becomes important to determine which pieces should be packed.

For example, when attempting to pack 150 g of green pepper into a bag, since the size and wight of the pieces of pepper varies, it would be extremely difficult to pack them into bags so that they weigh exactly 150 g.

Traditionally, creating a 150 g package involved a trial-and-error process that went on until the right combination of pieces was found. So when four or five pieces were packed into one bag, if the total weight fell short of 150 g, a small piece would be replaced by a larger one, and if the combined weight exceeded 150 g, a larger piece would be replaced by a smaller one.

Sometimes, there may not be a suitable piece of green pepper, so compromises were made and packages tended to be over the target weight to some extent. With such a method, work efficiency stagnated and the waste caused by the excess weights was not insubstantial.

Multi-head weigher

In response to an inquiry from an agricultural society on whether an automatic measuring system could be developed for the processing of green peppers into packages of uniform weight, a company solved the problem with a revolutionary way of thinking that gave rise to what is now known as the combination weighing technique. This company is *Ishida* Co. Ltd. (Sakyo-ku, Kyoto, President Takahide Ishida) and it developed the multi-head weigher, a computer-enhanced scale that can measure food items within a margin of error of 1g at a processing speed of more than 200 weighings per minute. The food items this scale can measure include even those whose individual sizes are large, such as vegetables and candies.[1,2] Established in 1893 in Kyoto (the 26th year of the Meiji era), *Ishida* is known for having been the first to develop the Temperature Compensated Spring Dial Scale and the Automatic Balance.

Combination Weighing Technique

As shown in Fig. 8.1, the combination weighing technique adopted by multi-head weighers involves the following steps:

1. *Subdivide*

In the multi-head weigher, there are around 14 containers called hoppers. Foods are subdivided and placed into these containers. The weights at this stage may be inconsistent. In fact, it is preferable for them to vary to some extent.

2. *Measure*

The weights of the food items placed into each container are weighed.

3. *Combine*

The weight combinations of the hoppers are considered. For example, in the case of 14 hoppers, their combinations will be $2^{14} - 1$ or 16383.

Fig. 8.1 Multihead weigher

Source: *Ishida*.

4. *Select*

Among the combinations mentioned above, the ones closest to the target weight are selected.

5. *Collect*

The selected hoppers are filled into bags.

6. *Refill the hoppers*

Food items are added again into the emptied hoppers.

The multi-head weigher carries out the above workflow for more than 200 times per minute. The most important step in the workflow is step number 4, "Select".

As shown in Fig. 8.2, food items of varying weights are stored inside each hopper so that combinations can be made to attempt to determine target weights. The combination patterns computed at this time are vast in number, and as shown in Fig. 8.3, it becomes possible to produce various weight totals.

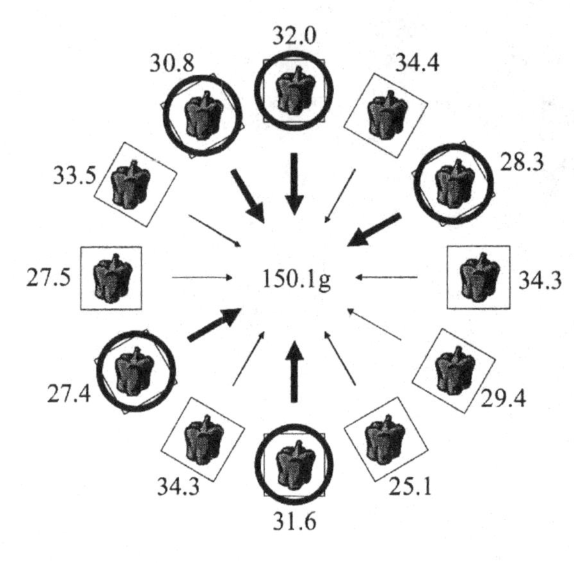

Combinations above 150g but closest to it are searched for and then filled into packers. In the case of 12 hoppers, 4,095 patterns exist.

The next food items are placed inside the emptied hoppers so that combinations closest to the target weight can be searched for again from the possible patterns.

Fig. 8.2 The line of thinking behind combination weighing

Source: Author.

As a result, it has become possible to determine target weights within a +0.5 g – 1.0 g margin of error. This is a surprisingly high level of precision.

The Algorithm Behind Calculating Combinations

The problem of this type of combinatorial computation, in the field of operations research, is similar to the problem known as "the knapsack problem".

The knapsack problem is the problem of maximizing a load to be packed under the constraint that the total weight of the load becomes less than or equal to a given limit. This problem can be understood to be the same problem combination weighing deals with, considering that it is about maximizing the weights remaining in the hoppers after subtracting weights that are less than or equal to the target weight from the total weight of the hoppers.

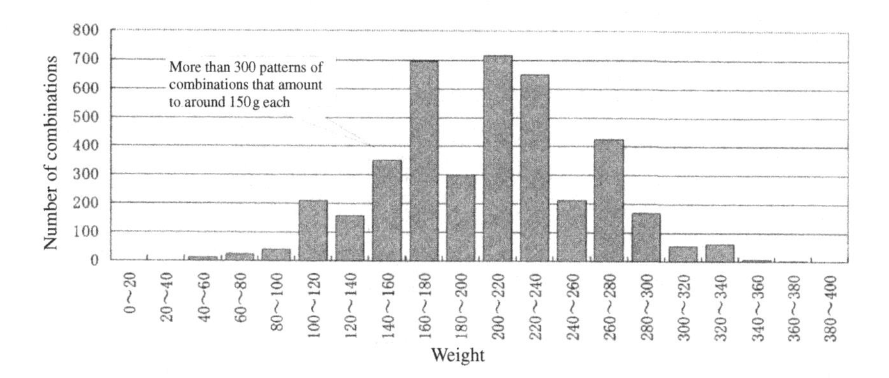

Fig. 8.3 Weight histograms obtained from combinations

Source: Author.

With respect to the knapsack problem, the algorithm for rapidly arriving at an optimum solution is known, so in the case of the combination patterns arising from 14 hoppers, the computation time required for determining all of them can be said to be brief.

The Spread of the Multi-Head Weigher

The multi-head weigher is adopted for special packaging of items such as asparagus, mini-tomatoes, and onions. In addition, another point of interest is the semi-automatic NFC series that processes slices of fish, which can spoil easily. Food items are placed on twelve plates, and in accordance with the predetermined target quantity and weight, the multi-head weigher indicates which trays have the foods that should be packed by turning on the lamps of the trays with such food items.

The operator packs these designated food items and when he or she replenishes the trays, the multi-head weigher repeats the workflow, going on to indicate the combinations of the next items. For the operator, the combination weighing technique allows him or her to weigh with a remarkable degree of accuracy, requiring him or her to only take measurements and then pack. In this way, the technique requires only a minimal level of human input. While its function is very simple, its effect is high.

Currently, *Ishida*'s multihead weigher sets the global standard in its field, enjoying a domestic market share of approximately 80% and a global market share of approximately 70%.

Bibliography

1. Ishida website http://www.ishida.co.jp/. Access on October 4, 2012.
2. Ishida, *Computer scale, Product information* http://www.ishida.co.jp/productlist/pro_list.php?cat2_id=1. (20121004)

9

The Remarkable Speeds of Sorting Machines

Takashi Yonezawa

Sorting Grains

The demand in Japan for food safety and quality is very high. Finding bad substances or foreign materials such as pebbles or grains of sand contained in foods is unacceptable. There is a company that possesses superior capability of detecting and removing foreign materials from a substantial amount of food.

Until several decades ago, it was not uncommon to see injuries caused by chewing a morsel of rice mixed with stones. To remove such a foreign material, people used to pass the grains of rice through a sieve or sort them according to weight by using a grain fan.

However, when the foreign materials could not be differentiated by size or weight, there was no other choice but to remove them manually with the aid of the naked eye. At present however, such removal of foreign materials has become mechanized and almost all grains of rice circulating and ending up on the dining table are treated by a sorting machine.

Anzai Manufacturing Co. Ltd. (Wakaba-ku, Chiba-shi, President Kenichi Anzai) was established in 1965 as a manufacturer specializing in the production of mechanized sorters for grains and thereafter went on to develop many technologies as the leading maker of sorting machines, as it continues to dominate in its field to this day.

In addition to sorters that sort colors, as shown in Fig. 9.1, there are also machines that sort out foreign materials, using something other than visible

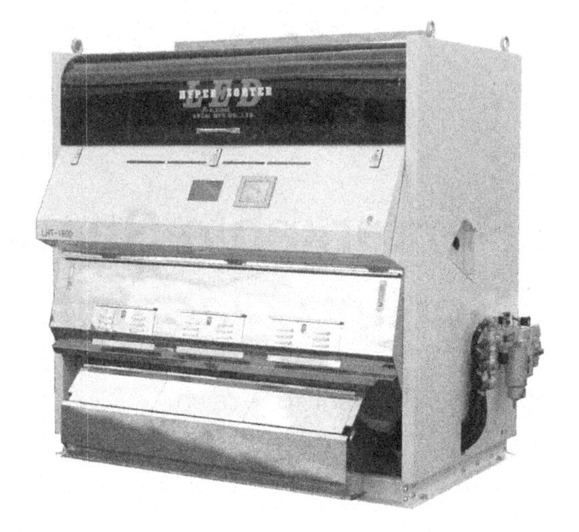

Fig. 9.1 A sorting machine for removing foreign materials

Source: *Anzai Manufacturing Co., Ltd.*[1]

light. Additionally, there are the chute-type models, which are suited for processing grains of rice and the belt-type models suited for more general use.

Colored/Foreign-Material Sorting Machines

The types of sorting machines available and the technologies they incorporate will be discussed below:

1. Color sorting machine

Fig. 9.2 shows the basic mechanism of the color sorting machine. A fixed quantity of food drops through a chute while a magnetic feeder makes it oscillate. Using two cameras or sensors to detect their colors, the machine distinguishes between non-conforming and conforming articles.

Foodstuffs determined to be non-conforming are sorted as such by being removed with an ejector.

2. Foreign-material sorting machine

Although the color sorting machine sorts by using simple colors or just three colors, it cannot distinguish glass or plastic pieces mixed in a quantity of polished rice.

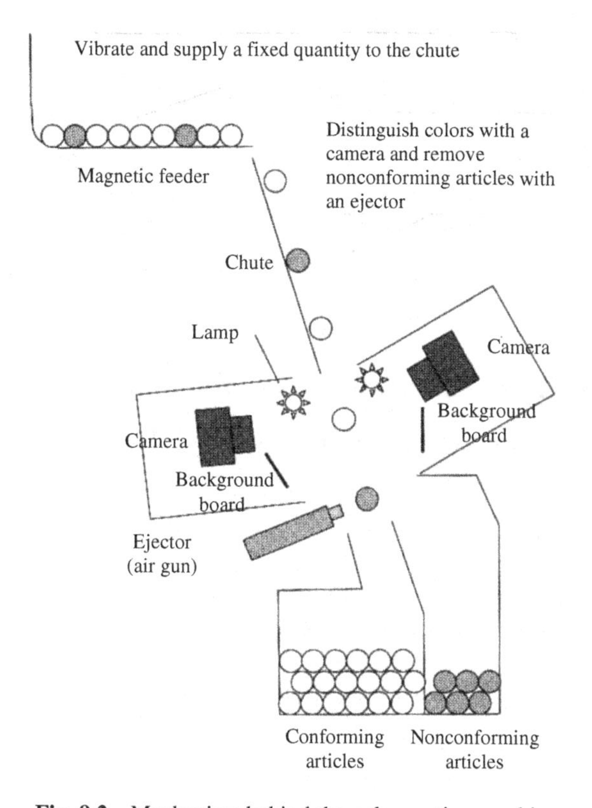

Fig. 9.2 Mechanism behind the color-sorting machine

Source: Created by author, based on the information contained in *Anzai Manufacturing's* website.

Consequently, the company has developed a model that can sort out foreign materials by using light that cannot be seen by the naked eye. Sorters in general sort out foreign materials by using near-infrared light, in the case of *Anzai Manufacturing*, in addition to such sorters, they use models that can sort with UV rays.

3. Chute-type sorting machine

For relatively friction-free, slippery rice grains of uniform size, the chute-type machine is adopted. The chute is a slide-shaped guide carved with 5–10 mm grooves, and the rice grains fall along these grooves. A single ejector (an air gun) is set into each groove, shooting the grains of rice falling from the grooves. A single unit of this ejector is called a channel and this unit forms the basic component of the sorting process.

4. Belt-type sorting machine

On the other hand, for foods that vary in size or for the purpose of distinguishing more general food items, the belt-type sorting machine is adopted. With the belt-type machine, the food is conveyed forcibly with the belt, and then, like the chute-type machine, it detects foreign materials and removes them with ejectors. In this case, unlike the chute-type machine, it does not have any grooves in particular, so it is necessary to have the food conveyed by the belt stably.

Sorting-Machine Performance

In terms of sorting-machine performance, the following three points are critical:

- Stable supply of desired item

Unlike the chute type model, it is important to control the flow of food so that it flows past the sensor at a fixed speed without piling up.

- Distinguish non-conforming articles (or rejects)

The most important constituent of the sorting machine is the screening component. At *Anzai Manufacturing*, even the lens, the Charge Coupled Device (CCD) sensor, and the circuit board have been developed in-house, which enables distinctive sorting speed and quality.

- Eliminating non-conforming articles

The ejector's performance and accuracy have been emphasized in the design, and the ejectors, which are placed at 5 mm intervals, are capable of removing nonconforming articles with pinpoint accuracy. In addition, their reaction speed and accuracy are exceptional, making it possible to eliminate almost all of the falling rice grains.

To shed light on these aspects, we shall examine the performance of the RSC-1800 model, which is considered to be the highest-end model in terms of performance. The RSC-1800 has the capacity to process polished rice grains at the rate of 18 /hour, and has six sets of 40 chutes, totaling 240 channels.

Since a grain of rice weighs 0.02 g, a single channel turns out to be sorting at a sensational speed of approximately 1,000 grains/second.

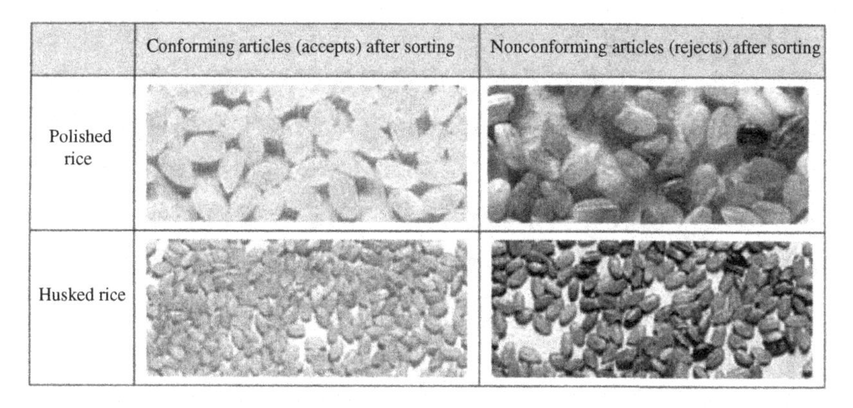

Fig. 9.3 Sorted samples

Source: *Anzai Manufacturing's* website.[2]

Similarly, an ejector can eliminate a grain of rice at this speed. In addition, multiple chutes can be used flexibly to increase the yield. To this end, the operator can set five chutes to carry out a primary sorting operation while setting aside one chute to specially sort out quality articles again from the non-conforming articles (rejects) that were sorted out in the primary sorting process.

The Development of the Sorting Machine

Fig. 9.3 shows actually sorted samples.

At present, there are models available for sorting not only cereals and pulses, but also various other product as follows.

- Other food items: Coffee, crepe, salt, potato chips
- Powders: Starch, parsley, green tea
- Industrial products: Pellets, ore

Bibliography

1. Anzai Manufacturing, *Sorting Principle, Products* http://www.anzai.co.jp/english/products/index.html. (20121004)
2. Anzai Manufacturing, *Sorting Samples, Products* http://www.anzai.co.jp/english/products/02_pro.html. (20121004)

10

A Ten-Fold Boost in Speed and Durability of the Zip Chain Lifter

Takashi Yonezawa

Lift Table (Lifter)

The lift table (a.k.a lifter) is a device utilized in the distribution and manufacturing sectors to transport people and objects up and down. There is an enterprise that has achieved more than a tenfold super effect boost in terms of both its speed and durability by applying an innovative technology. In general, the lift table commonly supports a table, on which people and freight can be placed, with a pantograph, which helps lift the table with a hydraulic cylinder attached to it. This is the common mechanism found in an ordinary lift table. By using the pantograph, it becomes possible to keep the height low during contraction while securing the desired height during extension. However, since the hydraulic cylinder gets pushed up diagonally during contraction (due to the structure of the pantograph), additional force becomes necessary, generating waste in the ascension rate as well.

Zip Chain Lifter

The Zip Chain Lifter created by the company, *Tsubakimoto Chain Co.* (Osaka-shi, Kita-ku, President and Representative Director Isamu Osa), as shown in Fig. 10.1, has challenged conventional thinking behind lift tables

Fig. 10.1 Zip Chain Lifter

Source: *Tsubakimoto Chain's* website.

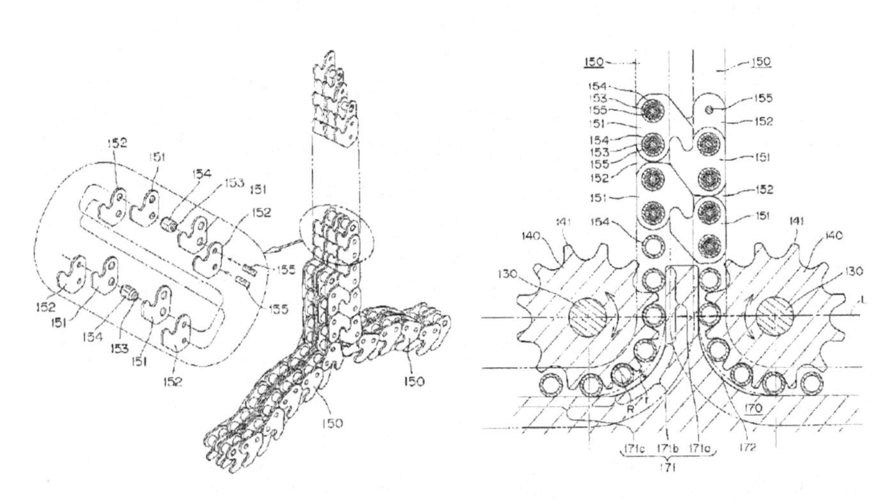

Fig. 10.2 The structure of the Zip Chain

Source: Patent Application No. 2009–132483 Figure 4, Figure 5.

and invented models with superior speed and durability. The chain was originally used for transmitting force by applying a pulling motion.[3,4] As typically demonstrated in bicycles and motorcycles, the key property of a chain lies in the fact that it is flexible and bendable yet can maintain a high level of tensile-strength.

Furthermore, it is lightweight, has a small footprint, and is also capable of transmitting force over a long distance.

The Zip Chain, as shown in Fig. 10.2, invalidates the basic mechanism of the chain, transmitting force *via* the application of a "pushing" action with the chain. Specifically, the Zip Chain works by having the teeth of chain mesh the way the teeth of zippers do, so that the chain becomes as stiff and unbendable as a metal rod, making it possible to push out. On the inner sides of two chains, there are mutually meshing teeth, and the wheel sprocket functions similarly to the slider of the zipper.

Comparisons with the Existing Technology

With conventional lift tables, as shown in Fig. 10.3, the table is lifted with an extending force applied on the link component of the pantograph by using a hydraulic cylinder. However, as can be seen in the case of the jack

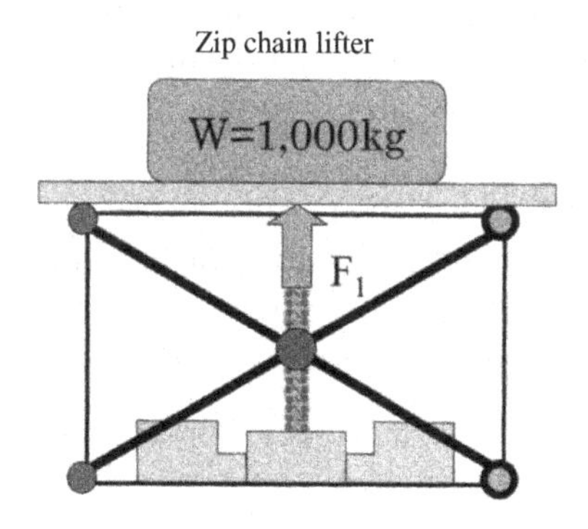

Only an unbalanced load weighs down on the hinge/roller component of the pantograph because the zip chain supports the weight.
F_1 = 1,000 kg
Speed: 0~50 m/minute
Durability: 1 million~1.5 million times

1/sin θ times the weight applies to the pantograph and hinge. The shorter the angle, the higher the load.
F_0 = 5.800 kg (θ = 10° → 5.8 times force)
Speed: 0~12 m/minute
Durability: 50,000 to 300,000 times

Fig. 10.3 Comparisons between the Zip Chain Lifter and the hydraulic lifter
Source: Adapted from data on *Tsubakimoto Chain's* website.[3]

used for cars, since the direction in which the hydraulic cylinder lengthens does not align with the direction in which the load-carrying platform is lifted in the beginning, it makes it necessary to apply a force of five to six times more than usual.

On the other hand, after the table lifts, since the directions coincide then, it becomes possible to directly lift the load carrying platform. Consequently, the telescopic motion (or the expansion and contraction) of the hydraulic jack fails to be proportional to the table's ascent and descent, making it difficult to control ascent and descent speeds with precision. Additionally, since the hydraulic pressure and the weight of loads directly apply to the hinge, problems of durability can easily develop.

With the Zip Chain Lifter, since the direction in which the Zip Chain elongates completely accords with the direction in which the load is pushed up, the lifter needs to only support the necessary load. In addition, since the table is fixed in place with the pantograph, the only force that can be applied to the Zip Chain is the up and down compressive force; no longitudinal force can be applied so no bending or twisting at the bottom occurs. On the other hand, where the pantograph itself is concerned, when a certain amount of weight is applied to the table, it only supports that unbalanced weight and no substantial amount of force is added.

With such a construction, the Zip Chain Lifter is capable of supporting a load of 5,000 kg at maximum, and it can be said that its durability is guaranteed, as demonstrated by the fact that this type of lifter has been adopted in the final assembly lines for the production of automobiles.

As for the speed of elevation, as shown in Fig. 10.4, compared to the speed of conventional lift tables, the Zip Chain Lifter's performance is far more superior, even in terms of its extension stroke or elevating length, which is more than twice as good. With regard to another similar technology, there is the spiral lift, but in terms of elevating speed, the zip chain enjoys a significant edge.[1]

Applications of the Zip Chain

A study was carried out at the Yokohama National University, focusing on a ship's stabilizer which leveraged on the Zip Chain's high-speed motion and the fact that it does not require a special mechanism on its upper part.[2] This is an example of the application of the Zip Chain. With this technology, the equipment possesses a staggering speed of 180 m/minute for sees raising and lowering a load weighing 180 kg.

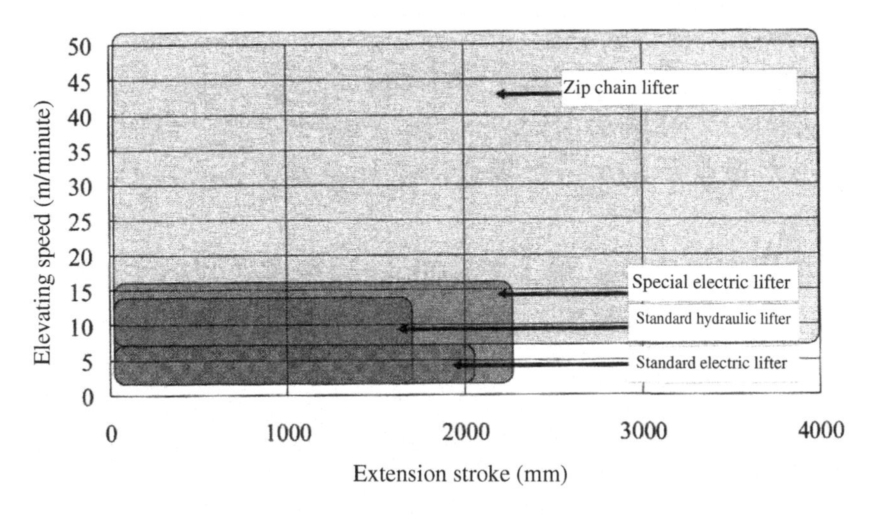

Fig. 10.4 Speed and stroke ranges of various lifters

Source: *Tsubakimoto Chain Co.'s* brochure.[4]

Bibliography

1. Akashin, *Spiral Electric Lift.* http://www.akashin.co.jp/slift.html. (20110606)
2. Mitsuharu Kakizoe, Tsugukiyo Hirayama, Yoshiaki Hirakawa, Takehiko Takayama, Naoki Okada, Akiko Yamane, *Development and Coastal Trial Runs of the Coriolis Vertical Anti-Rolling System for Ships, Journal of the Japan Society of Naval Architects and Ocean Engineers, The Japan Society of Naval Architects and Ocean Engineers*, No. 12, pp. 107–114.
3. Tsubakimoto Chain's website. http://tsubakimoto.com/. (20121004)
4. Tsubakimoto Chain, *"High-Speed Lift, Zip Chain Lifter, Products* http://www.tsubakimoto.jp/product/detail.html?code=450_1_1_1_1. (20121004).

11

The Introduction of The Virtual OS-Based Thin Client

Kazuo Matsude

Conventional Security Management for Networked PCs

Preventing the leakage of client information and mission-critical corporate data is now strongly recognized by many companies as a key issue. Most companies have their own in-house Local Area Network (LAN) for storing such information in a central server, allowing office personnel to access and process this information through PCs on a network. A survey of the patterns of information leakage originating from a PC revealed the following ways that leakages may occur: leakage *via* the Internet due to a computer virus; leakage *via* reproduction to a removable media such as a USB storage device or a CD; and leakage *via* a physical transfer of a PC unit (or hard disk) to an external location. The IT departments of companies are investing vast amounts of their resources in development and maintenance to have control over these routes.

Generally for most companies, as a rule, the practice of saving files to a hard disk is usually prohibited because doing so raises the risk of data leakage stemming from a computer virus or the loss or theft of a PC. It is desirable to restrict the operation of saving files itself through a computer program. However, imposing such a restriction uniformly could negatively impact the processing performance of other installed applications, so practically speaking, in most cases, companies cannot help but rely on the enforcement of a rule to restrict the operation.

Thin Client was introduced, which attracted attention with its claim to fundamentally resolve these problems by having hard disks disappear from networked terminals. The new network terminal that *Hitachi* Ltd. (Chiyoda-ku, Tokyo, Representative Executive Officer and President Hiroaki Nakanishi) released in January 2005 was based on such a concept.

Reassessment of the Thin Client and Its Issues

The emergence of the concept of the Thin Client dates back to the late 90s, but it is said that it initially attracted attention for its low cost and for its attractiveness as a way to extend the life of dated PCs. However, since PC prices began to drop rapidly, the Thin Client's price point was no longer considered to be that much of a merit.

The case of *Hitachi* Ltd. is noteworthy for having excavated new needs from the perspective of information security, which is a completely different concept. In other words, although the price merit of the Thin Client was, for the most part, lost, the introduction of the system came to be considered nonetheless by large departments in government authorities, financial institutions and firms in the communications service sector concerned with the social responsibility for maintaining proper information management. This is because the Thin Client made it possible to reduce that risk to zero in one fell swoop without relying on any elaborate enforcement of measures, thus achieving a super effect where the risk of information leakage due to the removal and transfer of hardware or infection from a computer virus was concerned.

The specification of the Thin Client that helped to realize the objective of information security is called the screen transmission mode. Most of the processing takes place on the server side and since the terminal side functions only as a monitor, apart from containing a lightweight Operating System (OS) to start up a flash memory, information doesn't remain in the terminal.

Depending on the server's type, the screen-transmission mode is further classified into the following three categories: the server-based type, the Blade PC type, and the virtual OS type. The server-based type involves a setup where all terminals share one server hardware, using software that supports multi-user access. The Blade PC setup is an extreme example of this type, and it sees the extension of the hardware to each terminal in the form of a blade, and on each blade, a user has access to the client OS and an ordinary single-user software. The virtual OS type is relatively new and belongs somewhere in the middle of the server-based category. In other words, this setup has a common server that runs a virtual OS that corresponds to the number of terminals connected to the server, and it has an OS environment for each terminal, running a single-user software on the virtual OS.

Fig. 11.1 shows comparisons between the conventional Network PC and various Thin Clients. The configuration of the server side of the server-based Thin Client closely resembles the conventional Network PC. However, since the server-based type cannot support system assets that do not support systems developed in-house, existing accounting systems, and multi-user access, it is necessary to consider other Thin Client types.

On the other hand, in the case of the Blade PC, the hardware corresponding to each individual terminal is installed physically as a blade, so there is no problem in handling software that is unable to run in a server-based setup. However, since the blade server is expensive, it fails to appeal on the grounds of price competitiveness, which is the Thin Client's original advantage.

The virtual OS type, as an alternative that compensates for the drawbacks of the two types mentioned above, enables the use of the single-user software, while aiming to maintain price competitiveness by avoiding the use of the blade server, making it ideal (Fig. 11.1). However, there are two conditions that need to be met for this model to become ideal not only on paper, but also in practice.

First, it needs to create a low-cost and reliable virtual OS environment on the server side. Secondly, it needs to be capable of incorporating Thin Clients on a large scale. The Silicon Valley enterprises, VMware and WYSE, which achieved sudden growth in the fields of virtualization and the Thin Client, appear to have succeeded in responding to these issues.

Further Super Effects

VMware has established a technology that stably and efficiently allocates virtual resource pools to every terminal by building the pools from a physical server. In addition, WYSE has a track record of stably operating a large-scale setup comprising more than 3,000 Thin Clients.

Thanks to the combination of these technologies, the time required for starting up a terminal has been shortened to approximately ten seconds.

	Conventional Network PC	Thin client (Screen-transmission mode)		
		Server-based type	Virtual OS	Blade PC
Server side	Multiuser applications Interface Server OS Hardware	Multiuser applications Interface Server OS Hardware	Applicat ion / Applicat ion / Applicat ion Virtual OS / Virtual OS / Virtual OS Server OS Hardware	Applica tion / Applic ation / Applica Terminal OS / Terminal OS / Terminal OS Hard ware / Hard ware / Hard ware
Terminal side	Applica tion / Applic ation / Applica tion Terminal OS / Terminal OS / Terminal OS Hard ware / Hard ware / Hard ware	Light weight OS / Light weight OS / Light weight OS	Light weight OS / Light weight OS / Light weight OS	Light weight OS / Light weight OS / Light weight OS

Fig. 11.1 Comparisons between the Network PCs and various types of Thin Clients

Source: Created by author.

This is made possible by the fact that the server is able to complete virus scans and update OS files during the night and from the fact that the Thin Client's OS is outstandingly lightweight. Since it used to take more than two to three minutes to start up conventional terminals, you can recognize a super effect here (in terms of time reduction).

In addition, due to virtualization, even the tasks of updating the OS and changing settings (such as the timer settings for the screen saver and screen lock) can now be carried out collectively in the central server — instead of in each and every terminal — promising drastic savings in the network's maintenance cost.

While the financial success of VMware and WYSE is greatly indebted directly to technology, it should be noted that this case example underscores the importance of simulation and innovation, the super effects of the virtual OS system.

Bibliography

1. *Hitachi, Ltd.*, Ubiquitous Platform Group, *Regarding the article, "Hitachi and the total abolition of personal computer usage," Hitachi, Ltd.* SEEDS Editorial Department, 2005. http://www.hitachi.co.jplProd/comp/OSD/seeds/521.htm. Accessed on April 13, 2011.
2. Bob O'Donnell, *Thin Clients: A Cost Effective Way to Improve Security,* IDC, 2004.
3. VMWare, *The Fundamentals of Virtualization, VMware,* 2011. http://vmware.com/jp/virtualization/what-is-virtualization.html. Accessed on April 24, 2011.
4. WYSE, *WYSE Thin Client,* Wyse Technology, 2011. http://www.wyse.co.jp/products/hardware/thinclients/. Accessed on March 29, 2011.
5. WYSE, *WYSE Technology User Casebook,* Wyse Technology, 2011. http://www.wyse.co.jp/resources/casestudies/. Accessed on April 23, 2011.
6. Masahide Ise, *Can the thin client replace the PC?,* Nikkei ITPro, 2008. http://itpro.nikkeibp.co.jp/article/COLUMN/20080327/297219/. Åccessed on April 23, 2011.
7. Masahide Ise, *The thin client is not almighty,* Nikkei ITPro, 2008. http://itpro.nikkeibp.co.jp/articlelCOLUMN/20080512/301373/. Accessed on April 23, 2011.

12

The Future of the Smart Meter

Kazuo Matsude

Installation of the Smart Meter

The smart meter is a next-generation watt-hour meter. The conventional watt-hour meter shows consumption with an analog display, and the meter reader takes readings from this meter once a month and reflects them in the electricity rate to be charged. In contrast, the smart meter makes a digital measurement of the amount of electricity consumed and transmits the watt consumption data to electric power companies and homes *via* a telecommunications facility. Since the interval between electric power measurements is every 15 to 30 minutes, a simple calculation shows the measurement frequency to be around a ratio to convention of 1,440–2,880 times every month (30-day conversion). The data becomes exceptionally more detailed and accurate, reflecting not only the number of monthly estimations of each household, but also fluctuations in electrical consumption registered in the middle of the month and during the daytime.

In addition, the smart meter is equipped with a controller that is compatible with the *ZigBee* standard for short-range wireless communication for consumer electronics. With this technology, the electric power company connects wirelessly to concentrators installed in telephone poles, making it possible for two-way communication with the electric power company *via* an Internet line, while also making it possible to control *ZigBee* supported devices in households and transmit watt consumption data to home PCs.

The watt-hour meter is a "specific measuring instrument" as stipulated by measurement regulations, making it necessary to acquire a seal of approval from the Japan Electric Meters Inspection Corporation. The period of validity for the seal of approval is 10 years and the electric power company is obligated to replace and carry out maintenance for all watt-meters for households at least once every 10 years. According to the data from the Ministry of Economy, Trade and Industry, since the number of watt-hour meters in use nationwide is approximately 74 million units, assuming that on average 10% will be replaced by smart meters every year from now on, a market of a little less than 150 billion yen a year will be created, assuming the unit cost of the meter to be 20,000 yen.

The makers of watt-hour meters are already focusing their attention on this future potential of the market. For example, the *Toko Electric Corporation* (Chiyoda-ku, Tokyo, President Funo Shunichi) and *Fuji Electric Co., Ltd.* (Shinagawa-ku, Tokyo, President and Representative Director Michihiro Kitazawa) have both started the development and sales of smart meters by establishing joint ventures of their own. In addition, with Toshiba Corporation (Minato-ku, Tokyo, President and CEO Norio Sasaki) announcing in May 2011 their acquisition of *Landis+Gyr*, the world's largest smart meter maker, for US$2,300 million dollars (186,300 million yen — at the exchange rate of 81 yen to a dollar), the move towards the smart meter is gaining momentum.

Merits of the Smart Meter

As the installers of the smart meter, electric power companies are the ones that stand to directly benefit from its spread. With the introduction of the meter, they can drastically reduce personnel expenses incurred for meter inspections. In addition, by leveraging the interactivity of the smart meter, it becomes possible for them to swiftly and flexibly carry out supply restrictions to non-paying customers.

The merit for customers, on the other hand, is higher clarity in understanding the breakdown of their electricity bill by being able to receive detailed electricity consumption data, allowing them to consciously change their behavior to reduce their consumption of electricity. What's more, it will become possible to set an alarm that goes off when the power consumption exceeds a fixed value. For example, *Google* equips smart meters that it supports in Germany with *Twitter* accounts to automatically transmit messages to cell phones that alert users when they are overusing electric power. The company provides this service for free.

In this way, with objects now that can connect to the Internet *via* a brain called the smart meter, it is expected that the Internet will see a new means of communication opening up in addition to the conventional modes of communication that see information flowing from "person to person" and from "person to object". Namely, this new mode of communication will see information flow from "object to person".

The Expansion of the Role of the Smart Meter — Moving Towards the Smart Grid

The flow of information originating from the smart meter promises not to be limited to the patterns mentioned above, but to extend to a pattern that will support "object to object" communication. The objective for this mode of communication, to be brought about by realizing an intricate electric supply and improvements in energy efficiency, is to keep wasteful power generation under check while achieving conservation of the global environment and the elimination of electricity shortage.

Consequently, what can be considered is to automate gates to efficiently load the power network with small forms of renewable energy that are locally produced for local consumption, such as wind power, photovoltaic power and biomass power. With the data from smart meters, the electricity consumption situation can be grasped on a region-specific basis in real time. So, during peak times, the gates of these power generation facilities can open to reinforce the power supply capability.

Alternatively, it is possible to consider plans for storing the surplus power arising from these power-generating facilities during off-peak times, such as during holidays and in the night. Consequently, storage facilities are deemed to be necessary, but technical problems still remain for large-capacity storage facilities. What comes to mind is the storage battery of the plug-in hybrid car and electric vehicle.

For example, to promote the spread of renewable energy and the electric vehicle in Denmark, IBM is developing a technology that distinguishes power for charging electric vehicles and synchronizes in real time the supply of energy derived from wind power generation. It is surmised that this technology's specifications will see the attachment of a *ZigBee*-compatible device to the charger plug of the electric vehicle. In view of the future prospects for the diffusion of plug-in vehicles, these storage facilities can be strongly expected to serve as buffers for the leveling of electricity consumption.

In this way, the basic concept of the smart grid (the next-generation grid) argues for a system that autonomously optimizes the supply/demand balance of electric power on a region-specific basis by dispersed processing without relying on the centralized control of the electric power company.

The Future of the Smart Meter

The impact of the rolling blackouts brought about by the aftermath of the 2011 Tohoku Earthquake and Tsunami disaster on Japan was extremely grave. The specific circumstances of individual regions inevitably became of low-priority due to the wide-area regional rotary shift system that was put into effect, exposing production lines and medical institutions to substantial risks.

Even in such an unexpected situation, however, if smart meters and a demand-response system are put together, risk reduction is considered possible. Zenichi Kishimoto, a representative of the U.S. firm, IP Devices, has pre-programmed the following demand response system made possible *via* smart meters and proposes to enter into contracts with electric power companies in the following way:

> "As ways to dampen demand, you can consider automatically raising the preset temperature of the air conditioner by sending a signal to the thermostat, turning off any unnecessary lighting, and turning off any other unnecessary apparatuses. So the proposal basically involves saving electricity by force, but it is better than carrying out a full-scale blackout through rolling blackouts".

It may be said that this is a new case example indicative of the future of the smart meter.

Bibliography

1. Ministry of Economy, Trade and Industry, *Summary of Regulations Affecting Specified Measuring Instruments,* Ministry of Economy, Trade and Industry, 2008. http://www.meti.go.jp/committee/materials/downloadfiles/g51216d06j.pdf. Accessed on April 30, 2011.

2. Norio Murakami, *New Internet Horizons Opened Up by the Smart Grid,* Gakushi Kaiho, Gakushikai, No. 887, 2011.
3. Hideyuki Kawai, *Impact of the Smart Grid and a Case Study of its Adoption Abroad — Overseas Trends and the Case of IBM's Adoption (August 2009 edition),* The Thirteenth Conference of the Japan Hydrogen Industrial Forum, 2009. http://www.ueri.co.jp/jhif/13Conference090828/C13_090828.html. Accessed on April 30, 2011.
4. Zenichi Kishimoto, *Rolling Blackouts are Drawing Attention to the Smart Grid,* Nikkei BP, 2011. http://itpro.nikkeibp.co.jp/article/COLUMN/20110405/359095/. Accessed on April 30, 2011.
5. Junichi Oba, *What is the Smart Meter? Why the smart meter is necessary for the smart grid and what is good about it,* TechnoAssociates Industry Innovation, 2011. http://sangyo.jp/ri/sg/na/article/20110408.html. Accessed on April 26, 2011.
6. Toshiba press release, *On the Launch of Toshiba Toko Meter Systems, Co., Ltd.,* Toshiba, 2009. http://www.toshiba.co.jp/about/press/2009_12_pr_j0101.htm. Accessed on March 28, 2011.
7. Toshiba press release, *On the share acquisition of Landis+Gyr — Accelerating the Global Expansion of the Smart Community Business,* Toshiba, 2011. http://www.toshiba.co.jp/about/press/2011_05/pr_j1902.htm. Accessed on May 29, 2011.
8. Japan IBM, *Introduction of European and American Case Examples for Steady Supply and Environmental Conservation — The Smart Meter and the Smart Grid,* IBM Business Consulting Services, 2007. http://www.meti.go.jp/committee/materials/downloadfiles/g71101c04j.pdf. Accessed on April 30, 2011.

13

The Acceleration of Processing Speeds Through the Evolution of Algorithms

Takashi Yonezawa

Linear Programming, Integer Programming

The technical innovation of the calculator has been remarkable, and needless to say, it has indeed produced super effects. Meanwhile, it is not well known that processing speeds have been seeing improvements as well *via* innovations in the programs used by computers, or in other words through the evolution of algorithms.

The *IBM ILOG CPLEX Solver*[2] of *IBM* (USA, Chairman of the Board Virginia M. Rometty) has achieved several-thousands fold of speed boost to exceed the calculator. But coupled with technological innovations of the computer, *IBM* has, in effect, achieved a several million-fold speed boost.

As a result, it has become possible to collectively optimize large-scale problems, which conventionally needed to be solved by breaking them down into smaller parts. In effect, it has become possible to consider adding more days to a planned schedule, which in turn is making it more possible to draw up a more practical and.

There is an academic field called Operations Research (OR). It is a study that carries out research on the methods of arriving at solutions for a business problem, based on the scientific method. One of its major themes is optimization, which entails presenting the most suitable combinations among other options after they meet given the conditions. The most important methods applied in this optimization are linear programming and integer programming.

For example, let us consider the case of delivering multiple products to multiple customers from multiple warehouses. Taking into account these warehouses' inventories, the quantities to be delivered to customers, the shipping costs incurred for the deliveries from the warehouses to the customers, what should the most efficient transportation volumes from each warehouse to each client be? This is known as the transportation problem and is one of the typical problems of the OR field, and linear programming and integer programming are applicable to such a problem.

The Basic Elements of Optimization

In optimization, the written expression of a real problem is based on the following three factors.

- *Decision variables*

Express a decision-making value as a variable. For example, from continuous variables such as amounts of production and transportation volume, there are also decision variables such as options like "will adopt/ will not adopt".

- *Constraints*

Determine the constraints that need to be met between decision variables. For example, the maximum capacity of a warehouse or the maximum capacity of a production facility becomes the constraint.

- *Objective functions*

The decision-making options determined by decision variables are expressed in a preference index as functions computed by the decision variables.

Linear Programming

Linear programming is an optimization method applicable when all the decision variables have continuous values. For example, the transportation volume and the amount of production are defined as variables that are continuous.

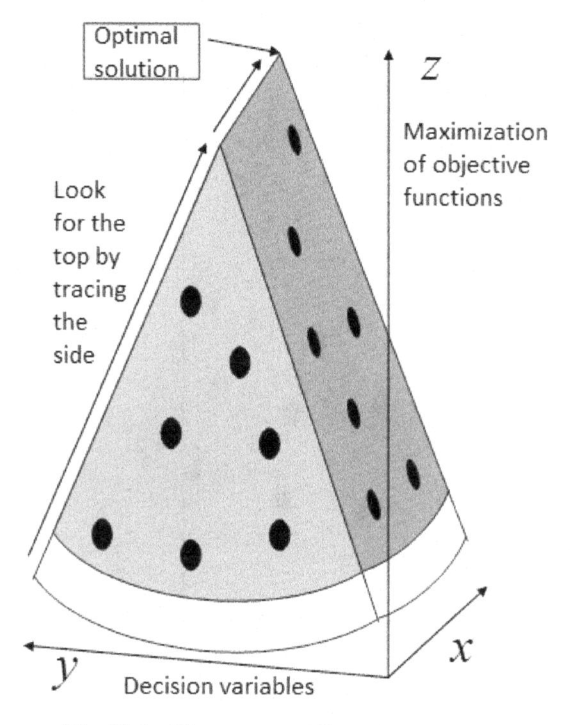

Fig. 13.1 The concept of linear programming

Source: Author.

In addition, constraints and objective functions are also expressed in the linear equations of decision variables. Upon meeting the requirements of these constraints, decision variables whose objective functions are optimal are sought, and decision-making is carried out on the basis of this result.

Fig. 13.1 shows this linear programming problem and the thinking behind the method of seeking the optimal values. So among the pulp inside the watermelon found on its cut surface (constraint), you would seek the position x, y (decision variable) that gives the highest z (objective function).

The most basic method is to trace an edge from its bottom to top, and keep repeating this until there are no more edges left to improve. In contrast, there is a method called the interior point method, which traces and searches the interior without relying on any edge. With these methods, it has become possible to seek optimal solutions within a realistic length of time, even for optimization problems that involve decision variables numbering in the tens of millions, due to various improvements.

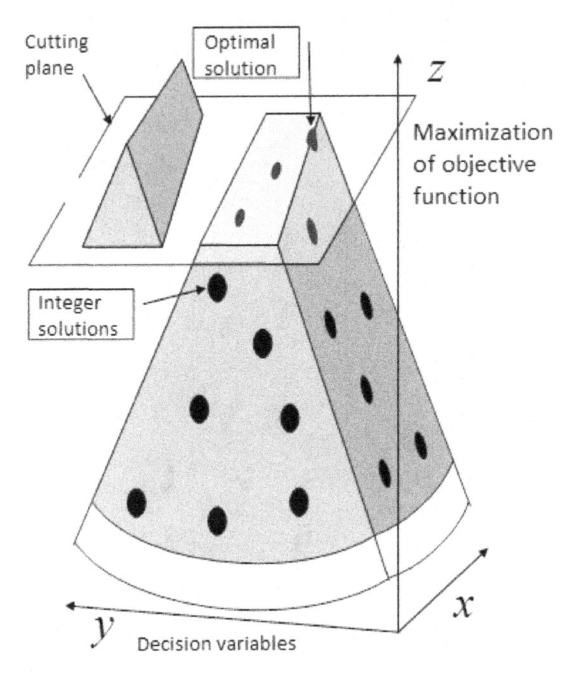

Fig. 13.2 The concept of integer programming

Source: Author.

Integer Programming

Integer programming is programming with the added constraint that a portion of the decision variables are integers.[1]

The fact that they are integers makes it possible to not only express things such as the number of vehicles or the number of people, which cannot be expressed in a continuous fashion, but also decision-making choices, such as "will carry out/will not carry out".

Fig. 13.2 shows this integer programming problem and the thinking behind optimal values. When decision variables are integers, the "seeds" are the solutions that meet the requirement of the constraint, and not the watermelon's pulp. Consequently, it is here that decision variable options with optimal objective functions are to be searched for.

Basically, it is a method to search exhaustively for all seeds (solutions that meet the integer constraint) with the branch and bound algorithm, and its key lies in how efficiently it is carried out.

A typical instance of this method is a method known as the cutting plane method, which facilitates and speeds up the search for seeds by efficiently cutting out only parts of the pulp in a way that allows the spots that meet the requirement of the integer constraint to come out to the surface.

Linear Programming, Integer Programming Solver

To apply linear and integer programming to real problems, it is necessary to express decision-making options and constraints, and objective functions in the formats of linear and integer programming. By processing them with a solver program, it becomes possible to seek decision-variables that optimize objective functions.

The solver program contains a protocol (an algorithm) that has been worked on in many ways to arrive at optimal solutions in a speedy manner, and the execution speed of the program is largely influenced by this algorithm.

A representative product of this solver program is *IBM's ILOG CPLEX*, which boasts a long tradition and track record.

Performance Boost and Case Study

A comparison between the *CPLEX V1.0* released in 1988 and the *CPLEX V8.0* released in 2002 reveals that the V8.0 achieved an 800-fold performance enhancement in terms of its computing power. However, the speed boost in throughput speeds achieved through algorithmic improvements saw a 2,360-fold increase.

Consequently, when we combine both effects, we see a performance enhancement of 1,900,000 times. Algorithmic acceleration is a result made possible by the achievements of many researchers. It is the product of the accumulation of all their achievements.

With the *CPLEX V12.0*, further speed boosts are provided, and when combining these boosts with the computing machine's performance, a several million-fold speed boost can be realized.

- Although integer programming differs from linear programming solely on the grounds that it adds the integer constraint to linear programming, by enabling multiple-choice decision making, its application has been extended to a wide variety of real-world problems. Concrete examples follow:

- Production planning
- Distribution plan

- Resource allocation plan
- Scheduling
- Two-dimensional cutting stock
- Timetable creation
- Personnel shift plan
- Advertisement allocation

Through optimizations of such actual problems, the *CPLEX*, with its superb processing speed, has made its application possible for large-scale problems and has made many achievements.

Bibliography

1. R. E. Bixby, M. Fenelon, Z. Gu, E. Rothber, R. Wunderling, Mixed-integer programming: a progress report, *The Sharpest Cut* (M. Grootschel ed.), SIAM, 2004, pp. 309–327.
2. IBM Corporation, *IBM ILOG CPLEX,* "Product" http://www-01.ibm.com/software/integration/optimization/cplex-optimization-studio/. Accessed on October 4, 2012.

14

Increasing Order Processing Capacity by 400 Times to 600 Times

Tetsuro Saisho

The Introduction of a New Concept

The case discussed here illustrates how building an information system involving a transaction process based on a new concept led to high-speed processing and an improvement in the level of convenience. It demonstrates the speeding up of order response times and information distribution by the newly constructed next-generation stock trading system in the securities exchange.

Amid the recent progress of the post-industrial information society and the advancement of financial engineering, the securities market is seeing an expansion triggered by the individual-investor driven growth of online trading and the introduction of program trading by institutional investors and securities companies. In such an environment of fluctuating money markets and diversification of transactional forms, the demand was rising for participating in a market environment where order response times and the speed of information distribution were accelerated.

At the Tokyo Stock Exchange (Chuo-ku, Tokyo, President & CEO Atsushi Saito), there are various market participants, and in response to a rise in each of their various needs, in order to strengthen its international competitive edge as a securities market, the exchange solicited ideas and cooperation from securities companies, information vendors, system vendors, individual investors, and institutional investors and went on to build a new system called *arrowhead* [sic].

arrowhead is a next-generation stock trading system founded on the concepts of low latency, high reliability, and scalability in the securities market, taking into consideration the specific conditions of the Japanese (Tokyo) marketplace, overseas market trends and the latest information technologies.[1]

The most significant feature of *arrowhead* is the low latency, or rapidity, of processing transactions related to orders. Due to this feature, processing times saw a drastic reduction, increasing capacity from the previous system's 400 times to 600 times. In addition, its design also takes into consideration the features of scalability and reliability.

Meanwhile, with the introduction of this new system, algorithm trading is expected to see an expansion. This form of transaction is becoming a global standard in the field of finance and it involves a computer carrying out automatic buying and selling based on the scenario of a portfolio assumed beforehand by the orderer.

However, those who are capable of handling such algorithmic trading are limited to entities such as major securities firms and foreign securities firms, and the environment this form of trading creates proves to be forbidding for individual investors and small- to medium-sized securities firms, who typically earn their profit margins through short-term buying and selling.

So in response, *arrowhead* created a world-class stock exchange system of the highest standard,[2] allowing faster order response times and information-distribution. Consequently, in the Tokyo Stock Exchange, through the operation of this new system, an improvement in the level of convenience for a wide variety of market participants was achieved.

Features of *arrowhead*

arrowhead [sic] is the name of the next-generation trading system that began operating on January 4, 2010 on the Tokyo Stock Exchange. It is marked by its three features of low latency, reliance, and scalability.[3] The transactions it handles are all auction trades that depend on spot financial instruments (equities and convertible bonds).

Auction trading is the method of transaction adopted by many securities exchanges in Japan and it provides a structure for validating trades when sell orders and buy orders, gathered in the marketplace, meet.

A survey of the three representative features of *arrowhead* will be discussed below.

The first is low latency, or speedy processing of transactions related to buy and sell orders. In effect, *arrowhead* has realized order-response times of five milliseconds and an information distribution speed of three milliseconds to enable access to the market at a millisecond level in terms of both buy-and-sell order processing and market-information delivery, which are key actions in a stock exchange. Again, that's access at the level of one millisecond, or one-1000th of a second.

The most substantial feature of *arrowhead* is its low latency, or rapidity, of processing transactions related to orders. In particular, the time it takes to process a transaction from the moment an order is placed to the completion of its processing is a mere 3mm/second. For example, by obtaining information on market trends, such as information on stock-price realizations or quotations, it becomes possible to carry out optimal trading behaviors. Optimal trading behaviors can tie into improving market liquidity and bringing about new transaction styles and business models.

The second is the securing of reliability to safely and confidently carrying out transactions related to buy and sell orders. *arrowhead* has been built to be a highly reliable system, leveraging the latest in information and communication technology. For example, the system has a structure that attaches great importance to preventing system failures, allowing for all market participants to access the information system confidently. In particular, this structure is brought about by ensuring that processing of transactional details related to orders, contracts, and order boards takes place on three-node servers. Ultimately, *arrowhead* supports not only the kind of processing that has low latency of the highest global standard, but by achieving a high level of trust, *arrowhead* also supports the Tokyo market as an information system where global market players transact.

In its primary site, it has realized a co-location service, which allows customers to set up their computers for automatic buying and selling at facilities well-equipped with the company's communications connections. Furthermore, with the creation of a secondary site serving as a backup center to alleviate the primary site's process load, it has made it possible to restore data within less than 24 hours even in situations of disasters affecting a wide area, substantially fulfilling its function as a societal infrastructure.

The third feature is scalability, which allows for the automatic interpretation of transactions related to buy and sell orders, and consequently extending the upper limits of threshold values. Currently, *arrowhead* continues to secure a capacity that is always twice the maximum processing capacity of actual conditions. Specifically, when the maximum system processing value under actual conditions for a whole day or the number of orders per minute reaches ½ of the predetermined system threshold value, the system automatically scales up the capacity to double the actual processing value. In addition, the processing for this scaling up is not only carried out on a daily basis, but also on a weekly basis.

Differences Between the Previous System and *arrowhead*

arrowhead is a next-generation system that reflects low latency, high reliability, and scalability in its design as shown in Fig. 14.1.

With regard to low latency, or rapidity, the capacity of the new system has seen a drastic improvement, equivalent to 400 times to 600 times the old system's capacity. For example, during trading, after matching individual issues, the securities exchange announces contract information and five quotations in real time. The previous system took several seconds to carry out a single match. In contrast, matching using *arrowhead* is immediate.

Fig. 14.1 *arrowhead* processing

Source: Author's photography.

With regard to reliability, *arrowhead* forecasts, monitors, and evaluates the resource demand sought after for IT systems and IT services from the perspective of capacity management, and plans, procures, and deploys to provide the system resources that meet this demand. For example, *arrowhead* calculates assumed values from the previous system's instances of orders, and based on market values forecasted for the period after *arrowhead*'s operation, *arrowhead* secures, at the time of operation, approximately four times the capacity of the peak values registered in the past.

As for scalability, the system flexibly improves its performance and functions, depending on the amount of increase in workload or the system's number of users. For example, with regard to the buying and selling regime, when *arrowhead* is in operation, it performs a review of various buying and selling procedures, such as reviewing distribution rules for simultaneous nominal quotations, discontinuing half-day trading while cutting back cuts on quoted prices, reviewing the price range limits and price range updates of quoted prices, and relaxing compliance requirements when determining opening prices and stop distribution actions.

Information from the Tokyo Stock Exchange released to individual investors is also spreading. For example, in addition to the number of fluctuations of buy and sell quotations increasing from five to eight, with the new service, FLEXFull, it has become possible to track, in real time, information on all orders across the board.

Bibliography

1. Nihon Keizai Shimbun, Tokyo Stock Exchange's high-speed buying and selling exceeds 1/3," May 31, 2011.
2. Yoshinobu Tanaka, *Tokyo Stock Exchange launces new stock trading system — order processing takes place at millisecond level, just as in the US and UK,* Don't stop business — ZDNet Japan, January 4, 2010. http://japan.zdnet.com/cio/analysis/20406097/. Accessed on September 30, 2012.
3. Tokyo Stock Exchange, *What is arrowhead?*, arrowhead Square http://www.tse.or.jp/rules/stock/arrowhead/info.html. Accessed on September 20, 2012.

15

A Ten-Fold Increase in Maximum Output of Photovoltaic Power Generation

Tetsuro Saisho

Industry-University Joint Research

The case discussed here shows how a new technology being developed through a research collaboration between the private and academic sectors is leading to the realization of a high-output power generation. Specifically, it is an account of a collaborative research being carried out by the University of Tokyo and *Sharp Corporation* with the aim to create a photovoltaic power plant capable of generating a high output of more than 1,000,000 kilowatts.

Photovoltaic power generation is a limitless electricity generating system for consuming energy from sunlight. Photovoltaic power generation is expected to be a new-generation energy resource, along with biomass power generation, solar thermal conversion power generation, snow-and-ice cryogenic power generation, geothermal power generation, and wind power generation because it does not release greenhouse gases and pollutants, such as carbon dioxide, methane, and the artificial material halocarbon, which are all said to be causes of global warming when carrying out thermal power generation.

At present, the largest output of photovoltaic power generation in the world is approximately 100,000 kW (100 megawatts) and it has been achieved by a photovoltaic power plant in Canada. Demand for the full-fledged utilization of the photovoltaic power plant exists in all parts of the world, particularly as a source of supply for meeting the large-scale demands coming from urban areas and industries. To produce power plants that can meet such demands, it has become necessary to establish photovoltaic power generation facilities capable of producing more than one million kW (1,000 megawatts), which is 10 times the maximum output of conventional photovoltaic power generation.

Under the aegis of the collaborative research being carried out by The University of Tokyo (Bunkyo-ku, Tokyo, President Junichi Hamada) and *Sharp Corporation* (Abeno-ku, Osaka-shi, Director, Chairman Mikio Katayama), the members of the project are planning to carry out a demonstration of establishing a photovoltaic power conversion technology and developing a 1,000,000 kW-scale photovoltaic power generation facility in Saudi Arabia by the end of 2014.

In the future, the project's plan also calls for proposing to foreign governments and overseas energy businesses the adoption of power supply *via* photovoltaic power generation and subsquently entering into contracts with them.[1]

Photovoltaic Power Generation and the *Status Quo* of Power Supply

In Japan, following the shutdown of Tokyo Electric's No. 1 nuclear power plant in Fukushima due to the accident that occurred in the wake of the Tohoku Earthquake and Tsunami disaster of March 11, 2011, then Prime Minister Naoto Kan demanded the shutdown of Chubu Electric Power's Hamaoka nuclear power plant. Consequently, an electricity shortage caused by the suspension of nuclear power plants exists today, causing a serious setback for modern society.

Aiming to resolve this issue of electricity shortage, municipalities such as Kanagawa prefecture (Governor Yuji Kuroiwa) and Shizuoka prefecture (Governor Heita Kawakatsu), in cooperation with Softbank Corporation (Minato-ku, Tokyo, President Masayoshi Son), are pushing forward with a plan to build large-scale photovoltaic power plants ("mega solars") all over Japan as a new-generation energy resource. In particular,

Overseas areas	Solar battery capacities	Features
Salzburg, Austria	200kW	Carrying out environmental measures by introducing photovoltaic systems in spaces with room to spare, such as airports.
Canary Islands, Spain	12.6MW	With a site area of 175,000m^2(approximately the size of four Tokyo Domes), this plant is the world's largest photovoltaic power plant backed by a Japanese company.
Sonnen, Germany	1.7MW	This photovoltaic power plant is a venture company financed by citizens, selling power to electric power companies and distributing the proceeds as dividends.
Tibet Autonomous Region	10.2kW	Installed a photovoltaic power generation system in a village that is 3,500m above see level without electricity in an interior part of China.
Qinghai Province, China	10.2kW	In China, the introduction of photovoltaic power generation systems in villages without electricity is being actively promoted.

Fig. 15.1 Large-scale photovoltaic power plants that have Sharp's participation

Source: Based on the information culled from Sharp's website.

the plan aims to bring forth the implementation of a full-scale means of photovoltaic power generation for general areas, promoting an alternative form of energy that will replace power generation by conventional thermal power plants that use fossil fuels, such as petroleum and coal.

However, for a full-scale photovoltaic power generation, a high output of at least 1,000,000 kW is essential, which is more than ten times the current output. As shown in Fig. 15.1, photovoltaic power generation makers, such as Sharp, Kyocera, Mitsubishi Corporation, and Sanyo Electric, are developing large-scale photovoltaic power plants abroad.

If Japan's photovoltaic power generation makers compare their technology to the technology of their overseas counterparts, they will see that Japanese makers excel in terms of generation efficiency and durability, and if high output power plants that leverage these advantages can be built, the international competitiveness of related industries within Japan will rise.

Therefore, the University of Tokyo and *Sharp Corporation*, as part of a collaborative effort between the private sector and academia for the achievement of building a high-output photovoltaic power plant, have joined forces with JGC Corporation (Yokohama-shi, Kanagawa, President Masahiko Yaegashi), J-POWER (Electric Power Development Co., Ltd) (Chuo-ku, Tokyo, President Masayoshi Kitamura), and the Development Bank of Japan (Chiyoda-ku, Tokyo, President Minoru Murofushi) to carry out with society an industry-university cooperation to establish a workshop concerning photovoltaic power generation.[2]

With respect to the practical use of high-output photovoltaic power generation, on the basis of their successful cases in Austria, Spain,

Germany, China, and Mongolia, they are building a high-performance 100,000 kW-class photovoltaic power plant in Saudi Arabia and conducting appraisals on various points, including the practical utility and cost of the technologies related to photovoltaic power generation. In Saudi Arabia, they are already in the process of selecting the site for the photovoltaic power plant, and are appealing to the national government for cooperation.[3]

Technology for the Practical Use of Photovoltaic Power Generation

With respect to the technology for practical use, in the joint research initiative being carried out by Sharp and the University of Tokyo, R&D is being conducted on how to create solar cells with more than 50% generating efficiency, which is double the performance of existing batteries; on the structure of supplementing climate-induced power variations by combining thermal power generation and storage batteries; and on cutting-edge electricity distribution systems that help minimize power transmission losses.

One of the results of these research projects is the successful raising of the generating-efficiency level to 42.1% with an apparatus called the Concentrated Photovoltaic Power Generation System, which collects light through a lens and irradiates it for a semiconductor that converts sunlight into electric power.[2]

The highest rate on record for conventional photovoltaic power generation was 41.6%, which was achieved by an American firm. But Sharp's technology, in effect, has achieved an efficiency that surpasses that record. It has been said that when generating-efficiency reaches a level above 45%, the power cost becomes equivalent to the power cost incurred by a thermal power station or nuclear power station run by Tokyo Electric's major electric power companies. At Sharp, experts are aiming to bring about products of mass production with even higher efficiency and durability by promoting efficiency of power generation based on further technical innovations.

The Concentrated Photovoltaic Power Generation System, as shown in Fig. 15.2, is suitable for generating photovoltaic power in vast locations such as the desert or farmland. While the technology of this solar battery is established, in practice, it is undergoing a testing

Fig. 15.2 Sharp's photovoltaic power generation (Spain)

Note: With the world's largest output of 12.6 MW, this photovoltaic power plant is representative of a Japanese enterprise's scope to serve as a project implementing body.

Source: Sharp's offical website.

regime that includes analysis of operative costs. Sharp is presently promoting research and development for further boosts in power generation efficiency and for realizing further mass production of the Concentrated Solar Cell System.

In the future, the aim is to supply to Europe, where there is a high level of environmental consciousness and where it is relatively easy to secure vast plots of land; to China, which is finally becoming active in taking measures to deal with the environmental problem; and to local and central governments, autonomous communities, and electric power companies overseas.

Incidentally, the scale of a 1,000,000 kW electric power generation is equivalent to the output made by a nuclear power station. However, since photovoltaic power generation is a form of electric power generation that is easily affected by the climate, its working efficiency will usually fall by a large margin during times of bad weather. With the current technology and performance, the working efficiency of photovoltaic power generation in Japan is ordinarily ten-odd percent, but in Saudi Arabia, where there is little rainfall, the working efficiency is expected to be several folds more than that.

Bibliography

1. Nihon Keizai Shimbun, *Tenfold Output Scale, Collaborative Research,* January 3, 2011.
2. Mainichi Shimbun, *42.1% Generating Efficiency — Sharp Becomes World's Best, Joint Development with The University of Tokyo,* September 7, 2010.
3. Sharp, *Industrial-use PV System* http://www.sharp.co.jp/solarsangyo/case. Accessed on September 30, 2012.

PART II

Case Studies of Global-Scale Super Effects

Section 1: Improvement in Sensitivity

16

Dramatic Changes in the Classification of World Competitiveness by Country

Akira Ishikawa

Declining Since 1993

When we rank comparisons by country, whether such comparisons relate to world competitiveness, an index of differences between men and women or average life expectancies, we need to take note of them in terms of the comprehensive results and index of new product/new business developments and improvements.

For example, with regard to the breakdown of world competitiveness by country, as shown in Fig. 16.1, the Swiss business school, the International Institute for Management Development (IMD), in their survey they have been carrying out every year since 1989, covering 55 countries and regions in the world, Japan ranked No. 1 in the world in 1991 and 1992, but since 1993, it has been falling, and in 1997 it suddenly dropped to 17th place. Since then, it has been floundering between roughly 20th and 30th place for these past ten years.

This world competitiveness comprises of the four categories: economic conditions, government efficiency, business efficiency, and the infrastructure. When we examine where Japan stands in terms of these rankings, we find that in terms of infrastructure, it is maintaining its rank in the top ten, at

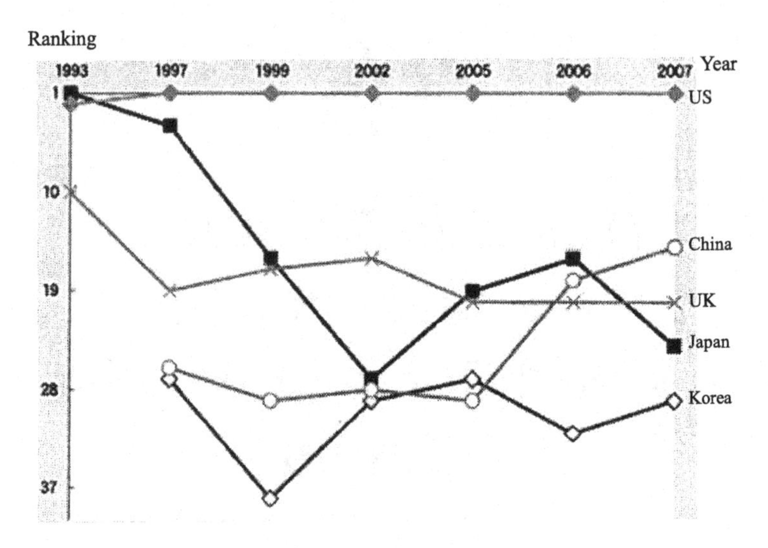

Fig. 16.1 Changes in world competitiveness

Source: The Ministry of Education, Culture, Sports, Science and Technology's graph based on the 2008 IMD World Competitiveness Yearbook[1].

sixth place. However, in terms of economic conditions, Japan ranks in 22nd. In terms of business efficiency, Japan ranks in 27th, while in terms of governmental efficiency, Japan is found at a pitiful 34th place.

Among developed countries, such a cataclysmic change in ranking is unparalleled. Meanwhile, with respect to rankings in the category of male and female differences, while they surely reflect the bias of societal structures and the degree of modern capitalism's development, where does Japan stand in terms of this ranking? The nonprofit organization, "World Economic Forum," (headquarters in Geneva) ranked Japan number 75 out of 134 countries in their report on male and female differences they release every year.

However, it is reported that the organization revised the ranking down to 101st place in the end of March 2010 after receiving further indications from organizations for Japanese women (April 5, 2010, Yomiuri Shimbun).

For a developed nation, the ranking was unusually low, yet it ended up dramatically plunging even lower.

With regard to the suicide rate and suicide total, which was raised as the third indicator, while the large number of suicides in Japan (over 30,000 people annually) is conspicuously reported every year within

Japan, according to WHO's rankings of suicide totals per 100,000 people among 106 countries, Japan is number four after Lithuania, Belarus, and Kazakhstan, confirming that Japan doesn't rank highly in this category. Incidentally, WHO predicts that the death toll caused by suicides will surpass 1,500,000 people in 2020.

With respect to military power rankings, according to the 2008 Jane's Defense Weekly, the US, France, Russia, China, and the UK are ranked among the top five, whereas Japan is ranked an anomalous 22nd. The major newspaper in India, Hindu, assesses Chinese military power to be at second place.

Seventh Place in The Human Development Index (HDI)

As for the Human Development Index (HDI), which Japan should take seriously, the United Nations has been publishing the report on it every five years since the year 2000. Japan's standing fluctuates between the 4th through 10th positions and in the 2006 report, it was not ranked at a high position, registering at 7th among 26 countries.

In contrast to such a low ranking, according to IMF's GDP World Economic Outlook Database for April 2010, the World Bank GDP 2008 and the GDP data from the CIA (Official Exchange Rate April 2010), Japan's GDP was 5,068.06, 4,909.27, and 5,108.00 (unit = one billion US dollars), which places the nation in third place after the European Union and the United States; and while Japan is predicted to surrender this position to China, Japan is maintaining a distinct lead over Germany in the fourth place and France in the fifth.

Impressive Average Life Expectancy

What Japan can be proud of, *vis-a-vis* other nations, is its average life expectancy. On the occasion of the 40th anniversary of the establishment of the medical department of Kitasato University held in June of 2011 at the Sagamiono Green Hall, Mamoru Mori, the astronaut and Director of the Miraikan (National Museum of Emerging Science and Innovation), delivered a commemorative lecture entitled, "The Connections of Life as Seen from Space". In this lecture, he emphasized that the longevity of the average lifetime is symbolic of a nation's level of progress as a civilization.

This is because, to maintain a high average life expectancy, it is not only indispensable to assure sufficiency of food, clothing and shelter, but also to see infant mortality rates decline; assure the availability of medical facilities, enhancements in nursing institutions and facilities for the elderly; and see improvements in innovations that elevate standards of living.

The audience received Mr. Mori well and gavehim a lasting standing ovation, convinced that he had addressed them with the intention to inspire the Japanese, a group of people who can become oddly submissive because they tend to be introverted and have an inferiority complex. In effect, he had reminded them of their nation's great achievement in the area of average life expectancy.

Bibliography

1. Ministry of Education, Culture, Sports, Science and Technology, *Intensification of World Competitiveness and the Need for Innovation,* 2008 edition — Technology White Paper, Ministry of Education, Culture, Sports, Science and Technology, 2009, pp. 20–24. http://www.mext.go.jp/b_menu/hakusho/html/hpaa200801/08060518/004.htm. Accessed on May 28, 2011.

17

The Spread of "Intelligence Olympics"

Akira Ishikawa

The Need for Intelligence Olympics

For humans to grow and mature holistically, a customary Olympics that attaches great importance to physical strength, or in other words, a venue where people can pit their physical strength against each other, is important. However, I have also been advocating, since more than fifty years ago, the need for a venue where people can pit their intellect and minds against each other, or in other words, an Intelligence Olympics, if you will.

This need came to be gradually recognized, and in 1950, as proposed by the Youth Career Association in Spain, the first International Skills Olympics was held in Spain (Madrid), in which contestants vied their intellectual skills against each other, and not their physical strengths. Since then, the event has been held in all parts of the world once every two years.

Furthermore, the International Mathematical Olympiad (IMO), which is included in the International Science Olympics, a contest associated with technology for students enrolled in the secondary curriculum (junior high and high school students) throughout the world, was held in 1959, and was followed by the International Physics Olympiad (IPhO) in 1967, which was in turn followed in 1968 by the International Chemistry Olympiad (IChO). And after a brief interval, the International Information Olympiad (IIO) was held in 1989, and in 1990, the International Biology Olympiad (IBO) followed.

Related tournaments in other fields were held as well; the International Philosophy Olympiad (IPO) in 1993, the International Astronomy Olympiad (IAO) in 1996, and the International Geography Olympiad (IGO) also in 1996, and the International Linguistics Olympiad (ILO) in 2003.

Japan's Lukewarm Support

When we examine what kinds of support are being carried out in Japan for these activities, we see that the Ministry of Education, Culture, Sports, Science and Technology is carrying out support through organizations for promoting technologies, which are independent administrative agencies, and through the Japan Science Foundation. However, it cannot be said that they are actively providing this support.

For example, Japan began participating in the International Mathematical Olympiad, the oldest contest of its kind running today, only since the 31st tournament of this Olympics held in Beijing, or 31 years after this Olympics began. Although Japan won five gold medals and one bronze medal in 2009 to become No. 2 for the first time, it has failed ever since then to enter into the top five.

Amid such a situation, it was difficult to simply remain a bystander, so Japan fulfilled the role of host nation in 2009 in the 20th International Biology Olympics, and even in the 42nd International Chemistry Olympics, which was held from July 19, 2010 through July 28, 2010, Japan was burdened with the heavy responsibility of serving as the host nation, despite being a newcomer, having participated from only 2003.

Japan Meets with Success Twice in the Theater Olympiad

Naturally, we can also include in Intelligence Olympics the Data Mining Olympiad, the Theater Olympiad, and even the enterprise-sponsored world tournament, Imagine Cup, which are all intended for a wider bracket. In particular, the Second Theater Olympiad held in Shizuoka contributed to the revitalization of the local community and is believed to have earned more than 230 million yen, which is considered extremely exceptional for this kind of an event (as of November 16, 1999). However, I would also like to touch on the Imagine Cup as a case study also illustrating exceptional, super effects.

Sponsored by Microsoft, this competition was first held in 2003 in Barcelona, Spain, and its mission was for Microsoft to support creative students who are driven to cause change in the world today through the use of technology. The event can be considered to be a global-scale technology contest.

Consequently, after its inception in 2003, the Imagine Cup grew into a global-scale event for searching solutions to real-world problems, and its Egypt convention held in 2009 turned out to be unprecedented in scale, drawing more than 300,000 students from more than 170 countries. The attendants include passionate and young programmers, mathematicians, engineers, designers, creators, and artists from around the world. They all prepared to compete in a competition category of their choice and to take up the challenge of changing the world.

What is noteworthy here is that the theme of the 2008 Imagine Cup, "solving societal problems with the use of technology," was adopted for the first time by the United Nations as the theme for the Millennium Development Goals since 2008.

The United Nations Millennium Development Goals comprised of eight goals:

- Eradication of extreme poverty and hunger,
- Achievement of universal primary education,
- Promotion of gender equality and the empowerment of women,
- Reduction of child mortality,
- Improvement of maternal health,
- Prevention of the spread of HIV/AIDS, malaria, and other diseases,
- Securing of environmental sustainability and,
- Development of a global partnership for development.

With 2015 as the target date, the goals entail substantial numerical targets to be achieved, as indicated in Fig. 17.1.

1. Relief for more than 500 million people sa id to be in the extreme poverty bracket
2. Relief for more than 300 million people suffering from hunger
3. Rapid improvements in the health of more than 30 million infants and more than 2 million pregnant women
4. The securing of safe drinking wa ter for more than 350 million people
5. Improvements in the state of hygiene for 650 million people
6. The acquisition of freedoms that are safer and based on more equal opportunities for more than 100 million women and girls

Fig. 17.1 Numerical targets

18

A 150-Fold Increase in Followers Through a Limited Discount

Junpei Nakagawa

A Pioneer in the Use of the *Twitter* Effect

With the restriction of the 140-character rule, *Twitter* makes it easier for users who are not much of letter writers to transmit information. It also helps them spread useful information they may find among the tweets of their followers with the re-tweet function (RT). For this reason and due to the fact that *Twitter* allows anyone access to any necessary information free of charge, *Twitter* has become one of the social media platforms to be brought into the limelight and praised for such usefulness.

Name registration-type social network services (SNS) that spread earlier, such as *Facebook*, which boasts the most number of users in the world, and mixi [sic], which has a large number of users within Japan, formed networks based on extremely close social relationships. In contrast, in *Twitter's* environment, there is no need for users to declare their real names and they can easily use handle names only. Consequently, both the transmitter and receiver of information can build casual relationships.

At this point, I would like to introduce the case study of the American firm, *Dell*, a firm that was the first to bring about a super effect with their adoption of *Twitter*, largely influencing other companies in the process.

This company established a *Twitter* account called *@DellOutlet* in 2007 as a tool to promote its outlet sales, which had failed to rise until then.

In the PC field, excessive competition is played out due to successive and simultaneous introductions of new products from companies, and it isn't infrequent to see troubled companies dispose inventories of outdated products. In addition, it is common in this field to see initially defective products, products damaged due to cargo collapses, and cancellations of reserved products. For PC makers therefore, to realize higher profitability, how they deal with their inventory of old goods while promoting their new products becomes vital.

Dell had already established in the PC marketplace the made-to-order mail order service for selling *Dell* models in which users familiar with PCs to a certain extent could appropriately choose their own specifications. This business model has been continuing since the time their sales were mainly carried out through mail order by phone, which is a long time before Internet mail order came into prominence. For this reason, *Dell*'s business model could be cited as a case that saw the establishment of its core competence — a type of customer support that could not be replicated by other companies — having established a competitive edge against new entrants into the personal computer mail order business through their in-house accumulation of exhaustive know-how.[1]

While *Dell*'s approach is very efficient in terms of avoiding the risk of returns from dealers and reducing the ratio of returned unsold goods, the problems of defective products and cancellations still persisted. Dell's outlet products refer to returned products that ship again after having been re-serviced in a factory. The major difference between outlet products and new products that require elaborate PR activities through the help of advertising agencies lies in the fact that, in the case of outlet products, too much time and budget cannot be allocated to support such advertising campaigns.

Consequently, *Dell* began to use *Twitter* from 2007, a time when the validity of marketing *via* social media was still an uncertainty, to consecutively carry out discount sales for users accessing the company's *Twitter* account. As shown in Fig. 18.1, *Dell*'s *Twitter* strategy turned out to be highly effective; a customer who made a purchase would tweet how much of a bargain he or she received by buying at the *Dell* outlet and the people reading that tweet would become followers of *Dell*'s *Twitter* account. Through such a cycle, after three years since launching its *Twitter* account, *Dell* achieved a sales figure of 6,500,000 dollars and a tremendous 150-fold increase in the number of its *Twitter* followers.

To attempt an improvement in a *Twitter*-driven sales performance, an increase in the number of followers — the people who read your *Twitter*

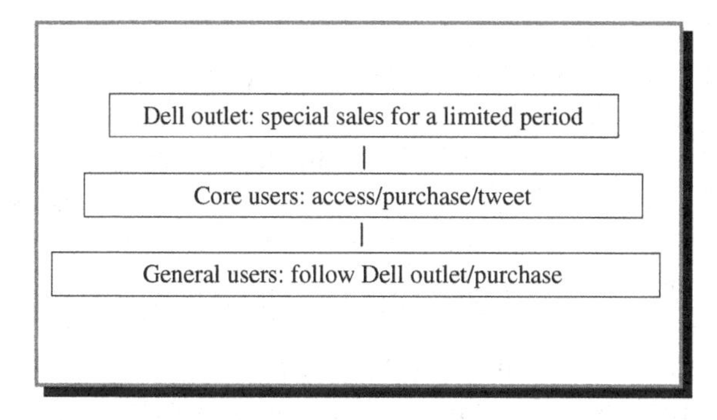

Fig. 18.1 Dell outlet's Twitter effect

Source: Author.

feeds — becomes a prerequisite. Even if information is transmitted to an unspecified number of readers, just as in a direct mail campaign, unless the recipients of the information become interested, that information will remain unread and will not readily tie into purchasing behavior. Although the Internet had become widespread, the age of e-commerce had arrived, and the medium had changed from conventional mail to electronic mail, making it easier to target information to even more potential readers, whether those readers would take an interest in the product remained an unknown factor, and it was difficult to boost the effect.

On the other hand, in the case of *Twitter*, since users access necessary information by only following specific accounts that interest them, the effect rises at an increasing tempo as the number of followers increases. It is this very aspect that gives a *Twitter*-driven marketing approach its super effect, and *Dell* Computer became aware of this insight quickly.

Moreover, *Dell* analyzed more than 12,000 opinions sent *via Twitter* and had made use of 350 of those views deemed to be useful.[2] The feature enabling interactive sharing of information with enthusiastic users is also an attribute of a marketing initiative using social media, and *Dell* has been leveraging this merit from an early stage.

Dell's Japanese Corporation also Begins Adoption of *Twitter*

For Dell, the trend of introducing *Twitter* has extended beyond the United States to Japan, where its Japanese corporation has acquired the account, "@DellCaresJP," to actively carry out information transmissions. Among

Japanese PC makers, some are major home-appliance makers and they tend to adopt a design philosophy that is user-friendly for novice users. As a result, for power users[a], the PC comes bundled with unnecessary software, creating a shortage in hard disk space when the PC's cost is already high to begin with. Consequently, the number of users who hesitate to purchase more domestic PCs is not insignificant.

Dell's business model, which does away with actual stores and enables users to customize according to their own specifications, was a perfect fit for the situation in Japan, helping the company win demand from the corporate market in particular, which comprised of users with a certain set of PC skills. Consequently, *Dell* went on to make rapid progress, surpassing Toshiba in 2010 to become No. 3 in market share.

The team behind *Dell*'s Japanese *Twitter* account is anticipating further share expansion by bringing in more first-time users. To this end, in an attempt to differentiate from their competitors, in addition to sharing "support information" with users in an interactive fashion, many of its tweets have included handy information on improving PC skills, such as tips on how to make a recovery disk for light users, how to extend memory capacity, and more recently, how to apply "energy saving measures" to deal with the problem of electric power demand. We can see *Dell*'s basic policy as a multinational enterprise reflected in the way its overseas branch has adapted their application of *Twitter* to meet specific local conditions.

Bibliography

1. Gary Hamel and C.K. Prahalad/Kazuo Ichijo, trans., *Competing for the Future (Japanese edition)*, Nihon Keizai Shimbun, Inc., 1994, p. 11.
2. Fumi Yamazaki, Koji Nozaki, Takuya Kawai, Toru Saito, *Twitter Marketing,* Impress Japan, 2009, p. 44.

[a]A power user is a user of a personal computer who has the ability to use advanced features of programs which are beyond the abilities of "normal" users.

19

Boosting Sales of Publications *via* Data Mining

Kazuo Matsude

Amazon.com's Super Effect

A Fortune 500 company, Amazon.com is an Internet mail-order behemoth based in Seattle, Washington, USA. According to its 2010 financial statements, the company's gross worldwide sales figure was 34, 204 million dollars. Of this amount generated, proceeds from book sales amounted to 14,888 million dollars (43.5% of gross sales). While the sales figures for the book sales of Amazon's Japanese branch (henceforth referred to as Amazon) were not disclosed, an explanatory footnote reveals that gross sales in Japan accounted for anywhere from 11% to 15% of the company's worldwide sales. Consequently, even a conservative estimate would suggest that Amazon's gross sales amount to 3,762 million dollars, or approximately 312,200 million yen (using an exchange rate of 83 yen to the dollar). Assuming the share of book sales to be 43.5%, which is the percentage for the company's worldwide book sales, Amazon's book sales in 2010 would have been 135,800 million yen.

Since even the highest sales figure among existing Japanese bookstores in Japan is Kinokuniya's 114,500 million yen (2009), Amazon remains the undisputed number one company in terms of book sales. Moreover, while Kinokuniya's sales declined 4.4% within the period from 2008 through 2009, Amazon continued to grow by a large margin, registering an exceptional growth rate of 27.9% in 2008 and 18.1% in 2009.[6]

The Secret Behind Amazon's Popularity

What is the secret of Amazon's growth and popularity? In my opinion, the secret lies in how the company makes you feel as if you have a smart personal librarian on tap just for you.

Firstly, Amazon stores records of purchases made for the last eight years. Secondly, the company also retains records of books checked but not purchased. And thirdly, the company "recommends" various books based on the trends revealed through these data points. Amazon, as a rule, retains data on purchase histories and browsing histories of all its customers, and also offers various proposals based on the data pertaining to other customers whose tastes are similar to this author's tastes. This process is known as data mining.

Shinya Ishikawa, a researcher at Sangaku Kyodo System Research Institute (Edogawa, Tokyo, President Tatsuro Shirai), defines data mining as "a sequential flow for excavating valuable information hidden inside data" and describes the characteristics of data mining, which differs from conventional statistical analysis, as follows.

1. A setup that can process large volumes of data
2. Capable of carrying out knowledge discovery without framing a hypothesis beforehand
3. With conventional statistical analysis, the approach taken is often one of summarizing multivariate data by reflecting them in the most effectively displayable linear fashion. In data mining, however, many of its techniques involve extracting specific pieces of data from an entirety, based on subject criteria.
4. A method for pre-editing data (cleansing) is established.

Amazon's Data Mining

Now, I would like to analyse the logic behind the step suggested by Amazon's "recommendations," as shown in Figs. 19.1 and 19.2.

Question: Customer No. 8888 just purchased product *A*. What will the "recommendation" for this customer be?

1. Collate sales data by customer number.
2. Extract data related to only those customers who purchased product *A* (cleansing).

Transaction No.	Customer No.	Product A	Product B	Product C	Date of purchase
100000001	1234	1	1		2011/1/1
100000002	3456			1	2011/1/2
100000003	7890	1			2011/1/3
100000004	7890		1		2011/1/4
100000005	1234			1	2011/1/5
100000006	2345		1		2011/1/6
100000007	3456		1		2011/1/7
100000008	2345			1	2011/1/8
100000009	4567				2011/1/9
100000010	4567	1	1		2011/1/10
100000011	8888	1	?	?	2011/1/11

Customer No.	Product A	Product B	Product C
1234	1	1	1
2345		1	1
3456		1	1
4567	1	1	
7890	1	1	
8888	1	?	?

Fig. 19.1 Customer-specific collation of data

Source: Author.

Customer No.	Product A	Product B	Product C
1234	1	1	1
2345		1	1
3456		1	1
4567	1	1	
7890	1	1	
8888	1	?	?

Customer No.	Product A	Product B	Product C
1234	1	1	1
4567	1	1	
7890	1	1	
TOTAL		3	1

Fig. 19.2 Among purchasers of product *A*, the number of purchases for other products (by product)

Source: Author.

3. Add up the quantity of purchases made for products *B* and *C*.
4. The "recommendation" will be the product with the larger total (= product *B*).

The examples above are extremely simplified and are displayed in an EXCEL format so as to be able to see how the data is processed. However, according to the actual data, the number of products processed is too enormous to be encompassed in a spreadsheet.

In actuality, the collation of sales data by customer number is not compiled into a single line data (since data length will vary by the number of transactions in each case), but is handled as multiple sets of sumset data by the customer.

Consequently, I gather that products other than *A* are tabulated after extracting all the records of the transaction associated with the customer who purchased product *A*. Just to be clear, Amazon has not disclosed this programming logic and, strictly speaking, this is my inference.

Issues Pertaining to Amazon's Business Model

When continuing to purchase books from Amazon in this way, you fill find that Amazon turns out to be very convenient since you receive more and more recommendations on reference works and related products that you were unaware of. However, this form of the company's retail marketing, which entails data mining, isn't without a few risks and problems of its own.

1. It is not possible to close personal accounts or erase personal purchase histories. Since personal accounts remain even after customers stop using the service, customers may express displeasure.

   Incidentally, companies handling private information in accordance with the Personal Information Protection Law are subject to restrictions that apply to deviation of original purpose or assignation of private information to the third party, but are not obligated to delete data unless the contents of the data are fraudulent. Since personal purchase histories are reflections of factual information, Amazon's measure to retain such records is legitimate.

   However, for a retail business, it is advisable to leave some room for complying with deletion requests in response to claims from concerned customers, even if such compliance comes in the form of offering the option to delete manually.
2. Marketing that has data mining as its objective must be carried out with a moderation that takes consumer protection into account.

   Formerly, the US's Amazon.com was criticized when it carried out a promotion that involved displaying varying bargain prices to different customers for the purpose of determining the retail prices of DVDs.

   In Japan, since "viable" prices set through such an experimental measure may not always be seen to be "fair," there is a risk that such marketing would end up receiving supervision from the Consumer Affairs Agency.

By responding to these issues in a sincere manner, the further growth of the online bookstore business can be anticipated.

Bibliography

1. Amazon.com, *2010 Annual Report,* Amazon.com, 2011.
2. Ian Ayres, *Super Crunchers: Why Thinking-by-Numbers is the New Way to be Smart,* Bentam Dell Publishing Group, 2007.
3. Todd R. Weiss, *Amazon.com Apologizes for Price Testing,* PCWorld, 2000. http://www.pcworld.com/article/18707/amazoncom_apologizes_for_price_testing.html Accessed on May 1, 2011.
4. Shinya Ishikawa, *The Treasure Chest of Data Mining,* Sangaku Kyodo System Research Institute, 2011. http://www.datamining.sakura.ne.jp/11haikei.html. Accessed on April 29, 2011.
5. Consumer Affairs Agency — Commercial Business and Price Regulation Division, *The first fact-finding committee minutes concerning the impact of electronic commerce on prices*, Consumer Affairs Agency, 2007. http://www.caa.go.jp/seikatsu/koukyou/data/18data/e-com190201-gijiroku.pdf. Accessed pm May 1, 2011.
6. Teikoku Databank Information Bureau, *The Publishing Trade's 2009 Statement of Accounts,* Teikoku Databank, 2010. http://www.tdb.co.jp/report/watching/press/pdf/p101101.pdf. Accessed May 1, 2011.

Section 2: Higher Precision

20

Customers' Return-to-Store Rate Through *Twitter* Usage Exceeds 50%

Junpei Nakagawa

The Food Service Industry's Shift in Strategy

The food service industry has had to endure severe conditions for many years now, having had to reduce material costs and personnel expenses to stay profitable as price competition intensified among both company stores and franchises. In particular, this trend has been remarkable among beef bowl and hamburger stores, and although such businesses have relied on the large scale of their operations, running large quantities of TV ads and many chain stores, they are supposedly mired in a state of store saturation.

On the other hand, mid-sized chain stores are aiming to increase customer traffic by distributing coupons through using "word-of-mouth" sites such as "*Gurunavi*" (Gourmet Navigator), but their efforts haven't readily led to any increase in the traffic of regular customers.[1] With the prevalence of competitors, the present state of the food service industry is not conducive to discovering effective measures, giving rise to the fear of the occurrence of the phenomenon of common ruin, which accompanies the intensification of price competition.

To overcome this problem, it is necessary to find a stable stream of revenue by building a strategy that completely diverges from those of competitors. To this end, if stores devise a method that can help pull in customers through utilizing *Twitter*, which has a high-level of local reach, even stores run by individuals will be able to secure a steady traffic of regular customers and avoid price competition with major restaurant chains.

As a case study to illustrate this, I will present the example of *Grace Inc.* (Minato-ku, Tokyo, President Jin Nakamura). This company runs the following four stores in Tokyo; the high-class pork-cutlet restaurant, "*Butagumi*", the *shabu-shabu* restaurant, "*Butagumi Shabuan*", and the store that triggered a boom in standup bars, "*Joe*". This company has a *Twitter* account and has already acquired nearly 10,000 followers, and 10% of their visitors are said to have found the store *via Twitter*. At this point, I would like to investigate the tips that small and intermediate sized stores could apply when they use *Twitter* to secure regular customers while avoiding competition with large enterprises.

Using the Official Account and the Personal Account for Different Purposes

In this case example, apt dissemination of information is carried out, thanks to the proper usage of the company's official account and personal account. Firstly, the official account, "Butagumi family (@butagumi)", mainly handles reservation from its *Twitter* followers and exchanges of after-meal impressions. The personal account is the account of the head of the company, "Jin Nakamura (@hitoshi)" and it works as a means to heighten customer loyalty through private exchanges with users.

What is unique about *Grace* is that it makes *Twitter*-mediated special offers not through its official company account, but through the president's private account. After building a close rapport between the president and his followers, there is a higher likelihood of seeing customers return to the store if the company carries out offers *via* the personal account.

For example, the company regularly carries out spontaneous campaigns, such as the "3 cm-thick Pork Cutlet Festival", a campaign that offers a free upgrade to super-thick slices of pork cutlets to visitors who are led to the store *via* the president's personal *Twitter* account, or "A Treasure Hunt Campaign", where visitors are offered discounts if they discover objects found in the photographs of the store's interior that had been uploaded on *Twitter*. As a result, 50% of visitors who are registered followers of the personal account become repeat visitors.[2]

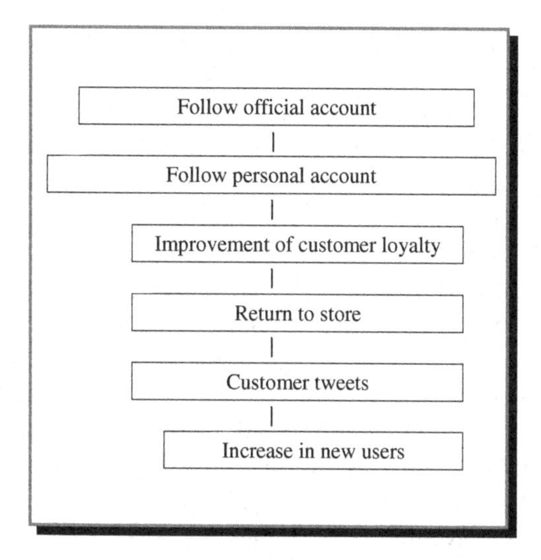

Fig. 20.1 *Grace*'s *Twitter* effect

Source: Author.

Merely lowering prices for a special time sale would not help the company to easily differentiate from other restaurants chains adopting a low-price strategy. The wiser policy would be to avoid cutting prices as much as possible and instead promote regular customers to return again by deepening their interest in the store.

Followers who discover information on an interesting event *via* the personal account make their reservations *via* the official account, and when they experience the event at the store, they tweet about it *via* their own accounts. As shown in Fig. 20.1, if followers of a customer who tweets about the event begin to become interested, a further increase in customer traffic can be anticipated.

Twitter as a Means to Implement a Blue Ocean Strategy

This case example is thought to be an exemplar of a "Blue Ocean Strategy", which has been drawing attention in recent years in the field of competitive strategy theory as it pertains to management strategy. As seen earlier, price competition in the food service industry has been gradually intensifying. While this means that consumers can make cheaper purchases, for sellers, this situation fails to lead to any growth in earnings, since even if they were to overcome their competition and realize an increase in sales, they would still have to contend with a sudden and cumulative rise in the cost of materials. This is an exemplar of a "Red Ocean Strategy", which is helpful in building competitive strategies for the purpose of defeating competitors in an existing industry structure.

However, in the case of *Grace*, the company went against the prevailing wisdom of the dining-out industry to position themselves as a business in the "knowledge industry" rather than in the conventional labor-intensive industry. Consequently, they made the decision to attach greater importance to making lifestyle recommendations to their visitors than offering meals and drinks to them. For example, by showing that drinking while standing is a lifestyle choice, "*Joe*" aims to become a standup bar for a new age. And by conveying "the richness of pork", "*Butagumi*" has been aiming to divest itself of the perception of being a B-class gourmet eatery (a type of restaurant known for cheap yet delicious food) on account of being a pork-cutlet restaurant.[3]

By acquiring regular customers *via* the use of *Twitter*, the company has succeeded in avoiding competition with existing bars and pork-cutlet restaurants. It may be said that their approach has been an exemplar of a "Blue Ocean Strategy", which is a strategy that aims to realize the acquisition of a stable and long-term revenue stream by creating "a new market that has yet to emerge", or by "digging up new demand", while raising value for customers.[4]

While Japan's economy is considered to have reached a stage of maturity, many of its industries have been experiencing excessive competition. Amid a situation that sees large enterprises attempting to survive by repeatedly carrying out mergers and acquisitions to form mega corporations, the road to survival for small- and intermediate-sized businesses may appear to be an extremely rough one these days. However, *Grace*'s approach, which entails the adept use of *Twitter* accounts that have a high-level of local reach, seems to be hinting that opportunities exist for small- and intermediate-sized businesses to discover their own competition-free "Blue Ocean Strategy" by engaging in a dialogue with their customers.

Bibliography

1. Gurunavi website, http://www.gnavi.co.jp/. Accessed on May 31, 2011.
2. Jin Nakamura, *The Case for Using Twitter to Help Small Stores Thrive — The Power of 140 Characters to Bond with Customers,* Nippon Jitsugyo Publishing, 2010, p. 32, pp. 70–71.
3. Grace Company Profile, http://www.grace.fm/profile/. Accessed on May 31, 2011.
4. W. Chan Kim, Renee Mauborgne/Yuko Aruga, trans., *Blue Ocean Strategy,* Random House Kodansha, 2005, p. 20.

21

3.5 Million Net Increase in Mobile Phone Contracts within a Year Due to *Twitter* Effect

Junpei Nakagawa

Reappraisal of the Growth Strategy — Improvement in Transmission Status

After *Softbank Corporation* (Minato-ku, Tokyo, President Masayoshi Son) acquired *Vodafone*'s Japanese legal entity with the key aim to promote its Internet strategy, *Softbank* went on the offensive in the mobile phone market dominated by *NTT docomo*.

Softbank Mobile, a mobile communications company, has helped introduce the mobile number portability system and launch a smartphone boom, and in 2010, the company achieved a net increase of 3.5 million phone contracts. However, one large factor behind this growth that cannot be overlooked is the fact that the company had implemented a *Twitter*-based interactive customer support program to improve the quality of connections, which had been considered a critical problem until then.

NTT docomo's predominance in the mobile communications business is unquestionably attributable to its ability to provide stable connections. To *docomo* users, this is a merit that justifies the company's comparatively higher call charges. Although *Softbank Mobile* had made strides soon after entering the market by deploying an extension strategy specifically aimed at the youth group, which included a discount service between subscribers, for future business applications, assuring stable connections had become a prerequisite. For this reason, to attempt a further expansion in market share, a review of their growth strategy was needed.

Top Management Recognizes the Necessity of *Twitter*

Softbank's President Son is also well known as a celebrity with a sizable *Twitter* following (@masason). When the Great East Japan Earthquake occurred in March 2011, he caused a sensation by tweeting his bold proposals for the adoption of alternative energy sources and his intention to donate substantially. In addition, in 2009, he had already embarked on a company-wide adoption of *Twitter*, which stemmed from his exchanges with his followers. While the total number of employees in the *Softbank* group is approximately 20,000, with the establishment of guidelines on social media use, almost all of these employees now have their own personal *Twitter* accounts.

In the wake of the Great East Japan Earthquake, blackouts occurred and caused a state of emergency in which TV reception became unavailable and both land line and mobile phone connections were cut off. Amid such a situation, people were able to send information in real-time from disaster areas *via* social media, including *Twitter*, and confirm vital information, including the conditions of public services and the safety of individuals, validating the power of this means of communication in the process.

However, in a good number of cases, some tweeters had re-tweeted (and helped spread) misleading information, since they had failed to check the accuracy of their data, causing anxiety in many recipients of the information. Social media allows the recipient of information to follow or not follow a sender of information, based on the recipient's perception of the sender's credibility, but at times of an emergency, since the sender is also in a state of confusion, a negative word-of-mouth effect could arise.

For this reason, a company is more likely to win a high level of trust from users if the sender of information in the company has received adequate training to be literate in the use of the medium and is able to convey precise information. If the sender continues to send reliable information and is able to become more trusted by his or her followers, then social media, which can cause further increases in the number of followers, can prove to be an extremely valuable tool for companies attempting to boost customer traffic and sales, proving its merit beyond its value as a tool for networking in the realm of social relations.

Whereas other companies had incorporated *Twitter* after their employees had voluntarily started to become active in the use of the medium, in the case of *Softbank Group*, the head of the company had recognized its necessity, and this ushered a new way of using *Twitter*, which goes beyond merely facilitating reciprocal relationships with customers.

As part of its investor relations program, the company has already been broadcasting its financial results briefings live since 2010 *via Ustream* and *Twitter*. A post *via Twitter* (a tweet) is distinctively casual, and in general, sending a tweet from the venue of an official occasion may harm an investor's trust in the company. But the tweets can only be distributed by the sender to avoid any misinterpretations of the details.

Conversion from Claim to Trust

Softbank Mobile Corporation, which is a mobile communications company, utilizes *Twitter* to carry out support for inquiries related to poor reception. *SB Care* (@SBCare), as it is known, is the largest customer support *Twitter* account in Japan at present and functions as a general customer support account. Furthermore, it is divided into the @SBCareDenpa account, which is in charge of the SBM area for 3G mobile phone lines and the @SBCareWiFi, which is in charge of wireless LANs, and carries out real-time responses pertaining to the reasons behind reception failures and on the status of future resolutions.

In addition, *Softbank* has also incorporated a service labeled "active support" to voluntarily search for tweets of complaints about reception and propose workarounds accordingly in the absence of direct inquiries from users with such complaints.

This customer service had already been carried out by the American cable television company, *Comcast*. The company distinguishes between "inbound" support, which is ordinary customer support that responds to inquiries from users, and "outbound" support, which involves voluntarily finding tweets related to complaints about the company's services. The company gives priority to the latter. In effect, the company has rapidly improved their customer satisfaction levels by assigning multiple accounts to its employees and having them search the *Twitter* sphere on a 24-hour basis.[1]

Softbank Mobile has also begun to carry out such advanced support duties through the use of *Twitter*. Although they don't operate on a 24-hour basis unlike *Comcast*, approximately 10 staff members are hard

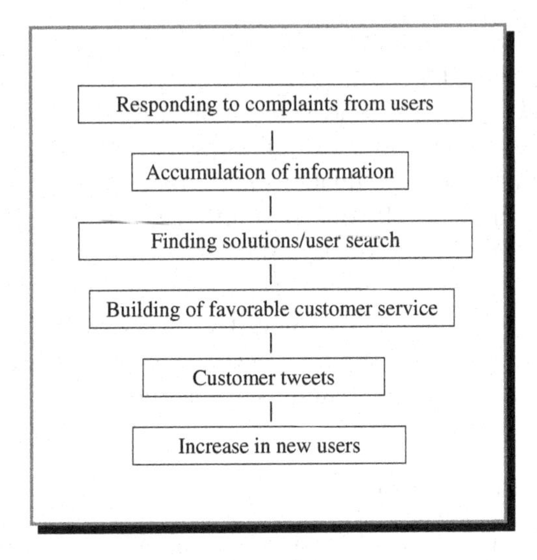

Fig. 21.1 *Softbank mobile*'s customer service

Source: Author.

at work everyday from morning to night, using more than 500 keywords for searching.[2]

With conventional support, the focus is on dealing with claims, and for this reason, most of the substance of this kind of support is negative. Consequently, for many companies, carrying out support is an obligatory exercise that requires enduring hardship. However, as shown in Fig. 21.1, when inquiries are made from the company side, users become surprised, but at the same time, they begin to express positive feelings for the company, which is when relationships based on mutual trust begin to be built in many cases.

In addition, if customer opinions could be compiled into a database, chances become higher that customer satisfaction levels will rise, as employees begin to share information among them for the purpose of improving communication quality and service.

Since the beginning of 2010, thanks to the boom of the *iPhone* sold by *Softbank Mobile*, smartphones began to rapidly spread in Japan as well, creating in the nation an environment that allowed easy access to *Twitter*. When this happened, the number of *Twitter* users continued to rise by more than 1.5 million users every month.

To deal with this rapid network expansion, the number of companies with *Twitter* accounts concomitantly rose rapidly, but the contribution of *Softbank Mobile* towards the construction of the infrastructure can be said to have been extremely significant.

Bibliography

1. Toru Saito, Looops Communications, Inc., *Social Media Dynamics,* Mainichi Communications, 2011, p. 25.
2. "In the looop [sic]" webpage, *Softbank SBcare, Japan's largest Twitter-based active support — Revelation of operation knowhow,* IT Media, February 25, 2011. Available at http://blogs.itmedia.co.jp/saito/2011/02/sbcare-0083.html (20110607).

22

Seawater Desalination Business Enabled by Reverse Osmosis Technology

Kazuo Matsude

The Technology Behind Seawater Desalination

With the advance of industrialization and urbanization in emerging economies, the demand for water is seeing a sudden increase, and in coastal areas devoid of fresh water sources, such as the coastal areas of the Middle East, the Mediterranean Sea, and parts of China, future growth in the demand for seawater desalination can be anticipated.

The technology behind desalinating seawater can be either classified into the evaporation method or the reverse osmosis method. Of the two, the evaporation method is the most traditional and it entails distilling water by applying heat to seawater. Its mechanism is simple, but its energy cost is huge. Middle Eastern oil producing nations have been conventionally investing their oil money into the establishment of many industrial plants that apply the evaporation method.

The reverse osmosis method, on the other hand, sets up a reverse osmotic membrane between seawater and fresh water, and by applying high pressure, exceeding the osmotic pressure, on the seawater side, moisture content in seawater is filtered out and transferred through the membrane to the fresh water side. At a glance, a sodium ion (about 0.13 nano in diameter: 1 nano equals 1/1000th of a micron) in diameter is smaller than a water molecule (about 0.38 nano in diameter), so it would seem likely that it would pass through the osmotic membrane to the fresh water side along with water molecules. But since the ion is hydrated and is linked to neighboring water molecules through hydrogen bonding, it becomes several times larger than a water molecule as a result, making it difficult to pass through the membrane. The aperture of the reverse osmotic membrane is 1 to 2 nano, which is one-tenth of the size of the aperture of a normal semi-permeable membrane used for filtration. Furthermore, the size of the aperture of a new reverse osmotic membrane that *Toray Industries, Inc.* (Chuo-ku, Tokyo, President Akihiro Nikkaku) announced in February 2011 is less than 1 nano by a large margin, placing it in a far superior league of ultra-thin precision. The contribution made by Japanese companies in this field can be said to be the technological harvest borne by years of engagement in the semiconductor domain.

However, with this method, it is necessary to add high pressure to the seawater side. In particular, the larger the difference in concentrations of salt in seawater and fresh water, the larger the osmotic pressure becomes. Since there is a tendency that water molecules would move in the direction of the seawater, pressure to overcome this tendency becomes necessary. According to the data of the product, *NTR-70SWC*, manufactured by *Nitto Denko Corp. Co., Ltd.* (Kita-ku, Osaka-shi, President Yukio Nagira), which has the top share in the market for reverse osmotic membranes for seawater desalination, the pressure required to lower the salinity level of seawater (approximately 3.5%) to less than 0.05%, which is suited for potability, is 5.49MPa (approximately a pressure of 54.2 standard atmosphere). However, even in light of this fact, the reverse osmosis method remains markedly cheaper to carry out than the evaporation method, so most large-scale industrial plants nowadays have adopted this method.

Favorable Business Prospects

Against the backdrop of economic development, the world water conversion business continues to grow at a rate of more than 10% a year, and the pie is certainly anticipated to expand, with the key momentum for growth coming from emerging nations.

In addition, for Japanese companies, which have the reverse osmosis technology, a technology that enjoys a comparative advantage relative to the evaporation technology in terms of cost, an expansion in market share, led by technology transfers during facility renewals, is also

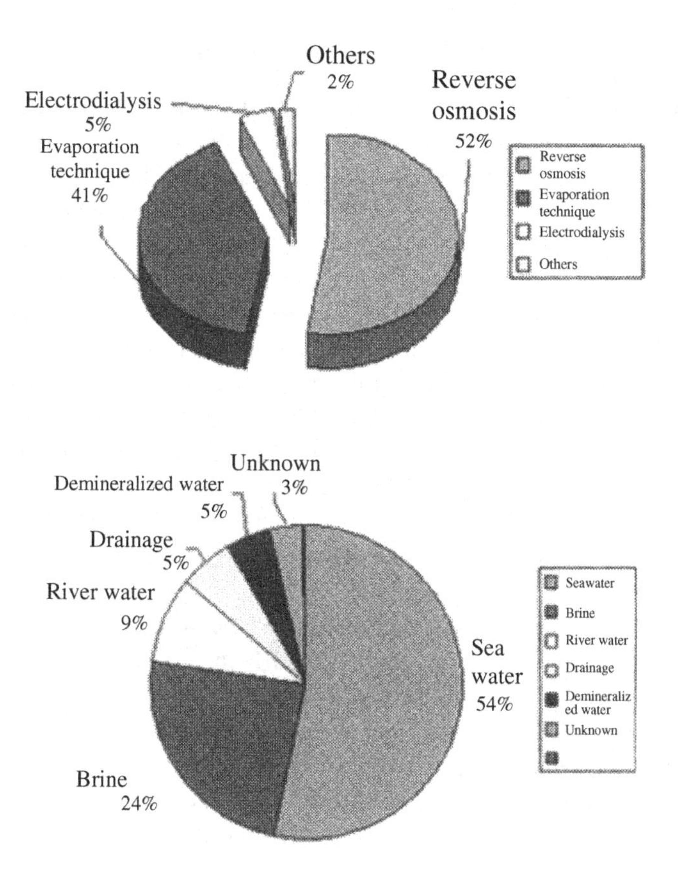

Fig. 22.1 Method-specific delivery achievements and raw water shares of desalination facilities around the world (2005)

Source: Mitsuyoshi Hirai, *The state of the diffusion of seawater desalination technology and its issues.*

anticipated. Fig. 22.1 shows the method-specific delivery achievements by water conversion facilities around the world, and chances are high that the 41% share of the evaporation method from now on will be replaced one after another by reverse osmosis. In addition, because reverse osmosis accounts for most desalination processes for river water and drainage, it is believed that in the end, decisions will more likely converge toward the adoption of the reverse osmosis technique regardless of the type of raw water used.

Based on the above-mentioned facts, it may be said that for producers of the reverse osmotic membrane, business is promising for the time being.

Issues of the Desalination Business and its Future

It would appear that Japan's desalination business is stable on the grounds of its membrane technology alone. However, this fact fails to reflect the total picture of the water business. In fact, large problems remain.

According to the Ministry of Economy, Trade and Industry's forecast, the market size of the world water business will reach 100 trillion yen in 2025, but of this, supply of membranous materials will account for only 1% with business administration and management accounting for 90%. Currently, companies that are dominant in such administration and management are all in Europe, where privatization of the water business got off to an early start. The so-called "water majors" control 80% of the global market: *Suez* (French) is in first place, *Veolla* (also French) is in second place, and *Thames Water* (British) is in third place.

To vie against these powers and enlarge the periphery of Japan's water business, what is first required is the cultivation of a leading figure. To this end, *Global Water Recycling and Reuse System Association Japan*, a limited liability partnership, was established in November 2008. With a membership of 50 companies in October 2010, the association declared its intent to carry out its activities with the aim to realize a system-wide competitive edge and to accumulate operational management know-how to incorporate comprehensively into a Japanese system.

What is required next is market selection. While in terms of the technology associated with reverse osmotic membranes, as mentioned previously, Japanese companies are the most advanced in the world. But it is apparently difficult for them to advance further and win contracts right away in the coastal areas of the Middle East, the Mediterranean, and China — areas where the water majors have already made their forays into. Apparently, the more rational step to take is to make forward investments in Southeast Asia (Malaysia, Thailand, and Indonesia) and India, which are economically growing regions and where the operations of the water majors are not full-fledged yet and where Japan has been contributing in the form of official development assistance (ODA) for developing countries.

In conclusion, due to the super effect of the core technology of reverse osmosis, the producers of raw materials can continue to enjoy favorable business prospects for the short term. However, if the public and private sectors were to join forces to reinforce the supply chain revolving around the raw materials, the producers can extend beyond their current level of success and anticipate further synergy. The chances for realizing this are by no means low if appropriate actions are taken strategically.

Bibliography

1. Nitto Denko Corp., Membrane Products, Product Data, *Seawater Desalination Spiral RO Membrane Element NTR-70SWC.* Available at http://www.nitto.co.jp/product/datasheet/membrane/008/index.html (20110504)
2. Mitsuyoshi Hirai, "*The State of the Diffusion of Seawater Desalination Technology and its Issues — Special feature: Water Resources,*" Science and Technology Agency, June 24, 2009, p. 3. Available at http://www.spc.jst.go.jp/hottopics/0907water/r0907_hirai.html (20110504)
3. Ministry of Economy, Trade and Industry, Industrial Science and Technology Policy Division of the Industrial Science and Technology Policy and Environment Bureau, Industrial Facilities Division of the Regional Economy, Trade and Industry Group/Planning Research Room of the International Trade Policy Bureau, *Our Nation's Water Business — Towards International Development of Water-Related Technologies, "Findings of the Water Resources Policy Research Association,"* Ministry of Economy, Trade and Industry, July 2008, p. 9. Available at http://www.meti.go.jp/policy/economy/gijutsu_kakushin/innovation_policy/pdf/mizuhoukokusyo.pdf (20110505)
4. Report on the promotional theme of 2007 by the Council on Competitiveness-Nippon, *Effective technological applications for Water Disposal and Water Resources — Approaches for the Fast Growing World Water Business Market,* Council on Competitiveness-Nippon, March 18, 2008, p. 4. Available at http://www.cocn.jp/common/pdf/mizu.pdf (20110505)
5. Global Water Recycling and Reuse System Association, Japan, *About GWRA. Available at* http://www.gwra.jp/jp/association/index.html (20110505)
6. Wataru Izumitani, *Reverse osmotic membrane, which transforms seawater into fresh water, sees a sudden rise — Technology cultivated in the semiconductor field heads for the environment,* Semiconportal, March 27, 2009. Available at http://www.semiconportal.com/archive/blog/insiders/izumiya/post-207.html (20110504)
7. Jiji Press, Zukai Shakai (Explanatory diagram society), "*Great East Japan Earthquake — Nuclear reactor recycling of polluted water,*" Jijicom, April 27, 2011.

8. Toray press release, *Seawater desalination plants in the coastal regions of the Arabian Gulf place successive orders for Toray's reverse osmotic membrane,* Toray, September 18, 2009. Available at http://www.toray.co.jp/news/water/nr080918.html (20110504)
9. Toray press release, *Success in developing a highly durable reverse osmotic membrane,* Toray, February 21, 2011. Available at http://www.toray.co.jp/news/water/nr110221.html (20110504)
10. Nozomi Nakai, *The Business of Fresh Water Generation — Reducing Global Water Stress for the Near Future,* Sumitomo Trust & Banking Monthly Survey Report, Sumitomo Trust & Banking, August 2009 edition, pp. 47–48. Available at http://www.smtb.jp/others/report/economy/stb/pdf/700.pdf (20120929)

23

The Surprising Properties of High Electric Current Density

Akira Ishikawa

The Roots of the Concept of Serendipity

In the well-known fairy tale titled, "The Three Princes of Serendip", the highlights are the chance encounters of the three princes, which lead to a series of other unexpected occurrences, which in turn lead to the inevitable discoveries of things the three princes weren't originally looking for. In essence, the story demonstrates the wisdom of being open to chance encounters.

The word serendipity is defined as "the natural ability to make valuable and interesting discoveries by chance (translation of the definition found in the Japanese version of the *Longman Dictionary of Contemporary English*). The term is often used in relation to great inventions and discoveries that suddenly appear one day amid unexpected circumstances and situations, as opposed to arising out of hardships and failures confronted over a lengthy period of time.

For example, in 1948, when George de Mestral was taking a walk along a country road in Switzerland, lots of thorny plants kept sticking on to his pants. Anyone could have such an experience, and upon removing these plants, the story would have simply ended there. However, George had carefully observed that the tips of the thorns were hook-like in shape and that they were caught in holes found between the weave textures. This turned out to be the spark that led to the discovery of *Velcro*.

Although such cases are too numerous to enumerate in their entirety, for this book, I would like to focus on the "carbon nanotube," a material discovered in 1991 by Professor Sumio Iijima, and touch on its raw-material properties and how it was discovered.

Serendipitous Discovery

The Professor himself describes the discovery of the carbon nanotube as being serendipitous indeed.[1] Stating that the principle of Archimedes, the discovery of Pluto, and the research findings of the Nobel laureates, Dr. Masatoshi Koshiba and Dr. Ryoji Noyori are all gifts of serendipity, he was apparently in the room of Meiji University's Professor Yoshinori Ando, when he caught sight of a heap of used up electrodes, which had been used to generate fullerene. Most people would surely assume the heap to be nothing more than just a pile of garbage, as it were.

The Professor, however, became curious by chance and observed the electrodes with an electron microscope, and in doing so, he discovered a long and thin carbon structure, which was inexplicable. In effect, in a completely unexpected way, a strange treasure had been concealed inside some garbage.

At present, the carbon nanotube is understood to a considerable degree; as shown in Fig. 23.1, this material is a six-membered ring network (graphene sheet) made of carbon and has the shape of a single-layered or a multi-layered (more than double-layered) co-axial tube. A carbonic allotrope, it is sometimes classified as a type of fullerene. It is important to note that the carbon nanotube is a carbon substrate whose discovery follows diamond, graphite, and fullerene.

The Surprising Properties of the Carbon Nanotube

For this reason, and in terms of its super effect, the carbon substrate has surprising properties.

For example, compared to copper, the carbon nanotube has a high electric current density that is 1,000 times greater, a high heat conductivity that is ten times greater, a high mechanical strength, and at half the weight of aluminum, it is lightweight. Furthermore, compared to steel, its strength

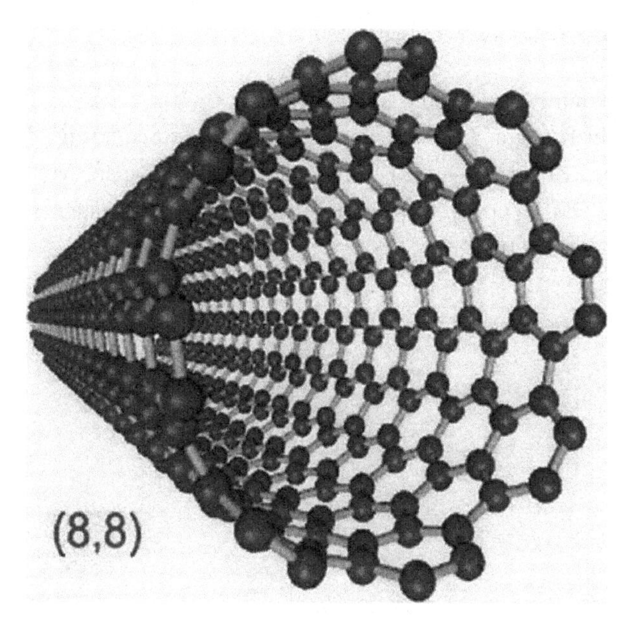

Fig. 23.1 (8, 8) A molecular model of a single-layered carbon nanotube
Source: Hiromichi Kataura, *Light absorption and Raman scattering of the carbon nanotube.*[2]

is 20 times greater; in particular, in terms of stretchability in the direction of its fibers, it is reported to surpass that of even diamond.

Moreover, since it has extremely flexible elasticity, it is anticipated to be used in the future as a raw ingredient of the rope to be used in the construction of the orbital elevator (a.k.a the space elevator).

As other remarkable structural properties of the single-layered carbon nanotube, experimental results have revealed that applying a pressure of more than 24GPa at normal temperatures to the material, using diamonds/anvil cells, can help the material to synthesize a tube that is super hard and has electrical conductivity; the volume modulus, as measured by the nano indenter hardness method of measurement, is 462-546GPa, surpassing the corresponding value for diamond, which is 420GPa.

But in the paper authored by researchers led by Professor Kenneth Donaldson at the University of Edinburgh and published on May 21, 2008 in the scientific journal "Nature Nanotechnology," the argument is made that products incorporating carbon nanotube technology have a high risk of causing lung cancer, since the carbon nanotube may give rise to health hazards similar to those brought about by asbestos.

Bibliography

1. Bungeishunju, April 2010 issue, pp. 84–86.
2. Hiromichi Kataura, *Light Absorption and Raman Scattering of the Carbon Nanotube,* Kogaku (Optics), Optical Society of Japan, Vol. 30, No. 2, February 2001, pp. 105–110. http://staff.aist.go.jp/h-kataura/Kogaku-kiji-forweb.htm (20110528)

24

A Fifteen-Fold Increase in Followers Through Ground-Breaking Activities

Junpei Nakagawa

Unusual Assortment of Goods and Marketing Activities

Tokyu Hands Inc. (Shibuya-ku, Tokyo, President Shinji Sakaki) was launched in 1976 as an urban DIY shop rooted in the rationale of "rediscovering the joys of making things with your own hands". Unlike other large stores that display popular items in large quantities, *Tokyu Hands* empowers its buyers more by offering a wider range of choices, diligently searching for unique products that consumers will want to buy. In this way, by attempting to stock its stores with a wide assortment of goods, the company has been carrying out its expansion in its unique way, increasing the number of its stores in and around urban areas.

Since goods at *Tokyu Hands* are not standard merchandises, mass advertisements for them cannot be expected to be sufficiently effective *vis-à-vis* their cost. For this reason, compared to those of other stores of the *Tokyu* group, *Tokyu Hands*' marketing activities have been consistently straightforward, never going beyond unique poster ad campaigns. For this reason, *Tokyu Hands* came to be widely perceived as a store for "people in the know" among those customers who would drop in one of its stores in the downtown areas of Shibuya, Shinjuku, or Ikebukuro by chance and discover the joy of browsing around and finding items unavailable at other stores.

However, with the spread of the Internet and strides made in search-engine accuracy, recognition of *Tokyu Hands* and its offerings gradually began to surge. In addition, since their stores were mainly located in the Tokyo metropolitan area and the company had not entered into local cities, to pursue further growth, it was necessary to increase mail-order sales. While the company achieved results to a certain extent through the establishments of its mail-order Web page, "*HANDS NET*," and its time-limited stores, what gave *Tokyu Hands* a chance to aim for a high-level sales effect was the appearance of social media such as *Twitter*.[1]

Twitter-Linked Site, "Korekamonet"

In March 2010, *Tokyu Hands* established the website, "korekamonet" (maybe-this-is-what-you-want.net) in cooperation with *Ryohin Keikaku*.[2] This is a service that was carried out as part of the Ministry of Economy, Trade and Industry's "undertaking to promote a new market created out of the fusion of IT and services," which is an enterprise with the aim to promote higher levels of productivity and innovation in the service industry. The *korekamonet* site is a service that is linked to the *Twitter* account, @korekamo, and it uses the recommendation and search engines developed by Team Lab Inc. (Bunkyo-ku, Tokyo, President Toshiyuki Inoko). On its opening day, the website was announced through tweets from the official accounts of *Tokyu Hands* and *Ryohin Keikaku*. These tweets drew in 1,000 users who became registered followers.

The service zeroes in on apparently appropriate products by using the *recommendbot*'s (recommendation bot's) natural language analysis. It works even when a user may not recall any concrete function or name of a product, that is to say even when the user has only a vague notion of what he or she is looking for. Specifically, by analyzing explanations and descriptions and matching them to the names of handled goods, the service conveys inventory information to the user. As shown in Fig. 24.1, the company, through the use of *Twitter*, is realizing with a high level of precision the essence of corporate marketing activities, which is converting potential needs to concrete wants.

As a result, in July 2010, *korekamonet* won the grand prix award in "The Second Nikkei Net Marketing Innovation Awards", sponsored by Nikkei BP. In having made the exchanges of questions and answers open for public viewing, the company was not only able to elevate purchase intention levels, but also trigger a viral word-of-mouth effect through the conceit of providing answers *via* a unique character.[3]

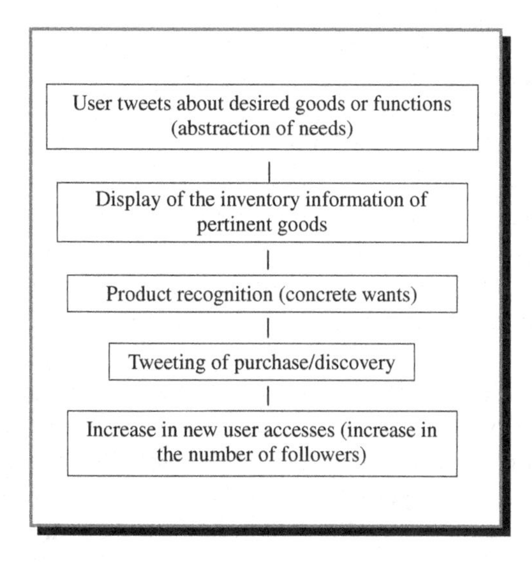

Fig. 24.1 Effectiveness of *korekamonet*

Source: Author.

As of May 2011, the company's number of followers exceeded 15,000, marking an approximately fifteen-fold increase in more than a year.

The Age of Marketing 3.0

This example of *Tokyu Hands* is thought to be a typical case of marketing 3.0, as advocated by marketing experts like Philip Kotler. Kotler positions product-centric marketing activities as "marketing 1.0". As illustrated in the case example of *Ryohin Keikaku*, this is typified by mass advertising-led marketing activities of the producer-centric age, and it is basically a "transaction-oriented" technique for handling inventories of mass-produced products.

Next, Kotler classifies marketing activities of the consumer-centric age as "marketing 2.0". This is a "relationship-oriented" technique that was implemented from the 1980s through the 2000s when marketing activities were subdivided. Specifically, it was when companies were pushing forward with customer segmentation and when word-of-mouth communication sites began to appear.

The modern marketing activities that are positioned as "marketing 3.0" are the ones that convey the mission of the company to each stakeholder, inviting consumers and ordinary citizens alike to cooperate in the development of products and in the co-creation of values. Kotler and other experts like him propose marketing activities that use social media. With regard to *Twitter*, they point out that the number of users increased 1,298% from April 2008 through April 2009 and attach great importance to the presence of users who praise and/or criticize companies and products through comments they tweet, which are far easier to make than blog entries since tweets can be easily sent out *via* mobile phones. In addition, they also point out that *Twitter* provides an opportunity to reexamine the relationship between customers and companies, citing cases that show how *Twitter* has been aiding marketing activities. In particular, with respect to companies, these cases show how younger employees have been teaching older employees how to use social media.[4]

In the case of *korekamonet*, although the website informs about goods that pertain to vague requests made by users, even with the unique and ample product lineup handled by *Tokyu Hands*, there are times when no appropriate items could be found. However, even in the event a product cannot be promptly offered, tweets regarding products sought by users are filled with plenty of insights on upgrading the store's assortment of goods, making *Twitter* something that the company has aimed to use; an objective means to deliver a service with a high level of reproducibility.

At present, *Tokyu Hands* cautiously assesses their use of social media to be experimental, since it has not led to any clear sales increases yet. However, for a company aiming to create a "hints market" — a reservoir of ideas on lifestyle tips from consumers — to provide consumers with a better assortment of goods in the future, *korekamonet* could be said to be a technique that is in line with the company's corporate philosophy.

Bibliography

1. Tokyu Hands, *HANDS NET.* Available athttp://www.hands-net.jp/shop/top/CSfTop.jsp (20110531)
2. Tokyu Hands, Ryouhin Keikaku, *korekamonet.* Available at http://korekamo.net (20110531)
3. Nikkei Net Marketing, *Nikkei Net Marketing Online,* Nikkei Business Publications, Inc. Available at http://business.nikkeibp.co.jp/nmg/special/2010_innovation_result/index.html (20110531)
4. Philip Kotler, Hermawan Kartajaya, Iwan Setiawan, Naoto Onzo and Kiyomi Fujii (translators), *"Kotler's Marketing 3.0,"* Asahi Shimbun Publications, 2010, pp. 22–23.

25

A Thirteen-Fold Increase in Registrants Through a Limited-Time Event

Junpei Nakagawa

Rapid Increase in Access Since the Inception of a Special Limited-Time Sale

Ryohin Keikaku Co., Ltd. (Toshima-ku, Tokyo, President Masaaki Kanai), is highly recognized as a company that offers products designed with simplicity and convenience in mind, having developed its original array of products under the brand name, "MUJI". As of May 2011, the number of followers for its *Twitter* account, @muji_net (MUJI.net), was approximately 136,200 people, making the account the largest of its kind in Japan at the time. In addition, among domestic companies, *Ryohin Keikaku* became a trailblazer in the full-fledged use of *Facebook*, and its number of fans there has exceeded 70,000.

A major factor behind the company's acquisition of many followers within a short time, as shown in Fig. 25.1, can be attributed to the fact that many of its followers and fans had re-tweeted the announcement of a special limited-time sale event, helping the information to become widely known among many *Twitter* users.

Specifically, to commemorate the fact that the number of followers of the company's *Twitter* account had reached 15,000 in January 2010, the company had decided to carry out a special limited-time sale and began to announce this decision one day prior to the event. In this announcement,

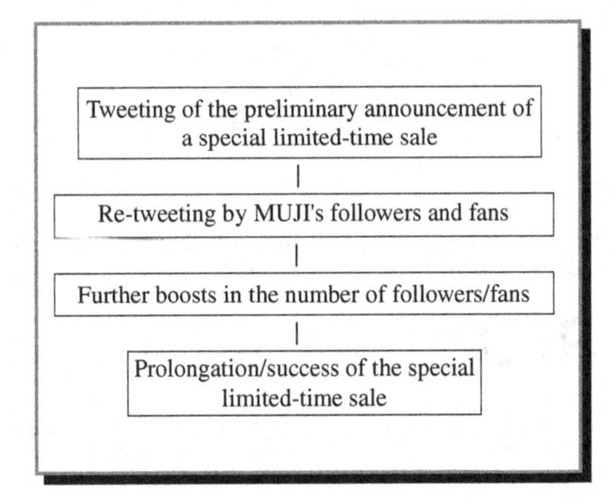

Fig. 25.1 *Ryohin Keikaku*'s *Twitter-Facebook* Effect

Source: Author.

the company tweeted that they will carry out the limited sale offer the next day, but the details would follow the next day, along with the disclosure of "a secret URL" to be revealed in the morning.

Later, as promised, the company tweeted, "Time sale now" in the morning and provided the URL of the limited-time sale site. *MUJI* tweeted intermittently to raise expectations and therefore succeeded in creating a high level of interest among its followers, who in turn went on to re-tweet instantly.

Because streams of *Twitter* users who found out about the limited-time sale began to follow the *MUJI* account and proceed to the special site to make purchases, *MUJI* repeatedly extended the sales period, and by the time they ended the sale, the company had achieved a favorable turnover of approximately 600,000 yen in four hours. Consequently, many reports on the special time sale began to appear on the Internet, stirring much interest among more users who went on to follow *MUJI*'s *Twitter* account one after another.[1]

As a result, the company now enjoys nine times greater the number of *Twitter followers* than prior to the limited-time sale event.

Facebook-Linked Product Planning

At this time, the company is contemplating the use of *Facebook*, where Japanese followers are still few, to carry out a product scheme that involves users. *MUJI*'s *Facebook* page is linked to the company's owned media, "*Kurashi no Ryohin Kenkyujo* (Ryohin Institute for Daily Living)," and shows that seven projects are underway as of May 2011.

For example, *MUJI* has developed the "*MUJI NOTEBOOK*" and "*MUJI CALENDAR*" applications for the *iPad* to allow users who have been using the company's stationery products for many years to use their digital versions on the *iPad*. However, some users said they would like *iPhone* versions of the apps as well, so the company went on to work on developing *iPhone* apps of the stationeries as well. But in this instance, the company decided to solicit opinions from users to help with the development of the apps.

Users immediately began to express their comments through *Facebook* and *Twitter* — comments such as "an extended zoom feature" and "improved recognition of handwriting input". The company has said it would refer to these comments to develop more convenient applications.[2]

Among Japanese companies, this was the first full-scale marketing initiative to utilize *Facebook* and it is spoken highly of by users. In January 2011, a limited-time sale was carried out to commemorate the fact that the total number of fans had exceeded 10,000, and in May, a gift-card giveaway campaign was carried out to commemorate having over 50,000 fans. For these instances, the scheme adopted required people to click on the "Like" button found on *MUJI*'s *Facebook* page to participate. As a result, the number of registrants exceeded 70,000.

The Cultural Basis of the Social Media Effect

Making full use of social media such as *Twitter* and *Facebook*, *MUJI* succeeded in increasing its number of enthusiastic followers and fans from 15,000 people to a total of 200,000 in less than one and a half years, achieving, in effect, more than a 13-fold increase. Another factor behind this achievement that cannot be overlooked is the fact that their corporate philosophy had been reflected in their social media campaigns; the philosophy of promoting consumers to change their lifestyle from an ostentatious, brand-driven consumption one to a high-quality and simple one.

Ryohin Keikaku (*MUJI*) was established at the time when the company's marketing activities were reaching a turning point. Since the 1930s, when full-scale marketing activities began in the United States, companies had been having the upper hand. But this situation lasted only until around the 1960s, after which consumers began to gradually become a countervailing power against the companies.

While GM spent on marketing activities, they were downplaying the importance of investing in automobile safety. The consumer activism led by Ralph Nader, which saw him rallying against this neglect by trying to acquire equity in the company to gain the right to speak at the company's general meeting of stockholders, is a representative example of the countervailing power. In the case of Japan, there was the famous "30-year war" that saw *Matsushita Electric Industrial* (currently *Panasonic Co., Ltd.*, Kadomashi, Osaka, President Fumio Otsubo) and *Daiei Inc.* (Koto-ku, Tokyo, President Michio Kuwahara) vie against each other over selling prices.[1]*

MUJI started in 1980 as the private brand of *Seiyu* (an affiliate of *Wal-Mart* Stores, Kita-ku, Tokyo, Steve Dacus, CEO). The brand was founded as an anti-establishment alternative that values consumer sovereignty and opposes corporate marketing activities that attach great importance to branding that makes full use of the mass media. In essence, it aims to help consumers feel free from being conscious about the brand name or the company's name of the products they consume.[3] *MUJI* has been selling standard mass-produced products at low prices and adopting an approach that completely differs from *Daiei*'s successful sales strategy to foster the emergence of enthusiastic customers by continuing to offer simpler products since the 80s, when lifestyle changes were being promoted; namely people were being encouraged to adopt more individualistic consumption as opposed to a corporate-led one.

What should be pointed out here is that in the case of *MUJI*, when social media emerged, the groundwork had already been established for users to be able to directly exchange information with companies and foster a simple lifestyle together. In other words, not every company can immediately achieve a significant effect by just using *Twitter* and *Facebook*.

There have been many cases whereby companies set up their own accounts in response to the trendy phenomenon of *Twitter* only to see a drop in the number of followers or fans because they failed to grasp how to use the medium effectively. For companies, what will determine whether the effectiveness of their use of social media can be heightened is whether they are offering products or services that require interaction with users.

Note

*In 1964, *Daiei* sold *Matsushita Electric*'s products at a 20% discount. However, since this basically turned out to undercut the implicit agreement for the discount range with *Matsushita*'s company stores — which was up to 15% — Konosuke Matsushita stopped shipping to *Daiei*. Consequently, *Daiei*'s President, Isao Nakauchi, brought an action against *Matsushita Electric* for violating the anti-trust law and also helped consumers boycott *Matsushita*'s products. Eventually, *Daiei* won.

Bibliography

1. TECH WAVE, *Time Sale Now = MUJI achieves great success with Twitter.* Available at http://techwave.jp/archives/51411971.html (20110531)
2. Kurashi no Ryohin Kenkyujo, *iPhone application.* Available at http://www.muji.net/lab/project/app/ (20110531)
3. Seiji Tsutsumi, Akira Miura, *MUJI Nippon,* Chuko Shinsho, 2009, pp. 97–100.

Section 3: Increases in Effectiveness and Efficiency

26

CO_2 Reduction Business Utilizing Electric Hydraulic Shovels

Toru Fujii

Summary of the Business Involved in Reducing CO_2 Emissions

Hitachi Construction Machinery Co., Ltd. (Bunkyo-ku, Tokyo, President Yuichi Tsujimoto) announced that it will set up a business that specializes in reducing CO_2 emissions with the use of an electrically driven hydraulic shovel, together with the *Ishizaka-Group* (Iruma-gun, Saitama, President Noriko Unemoto), a waste management company.

This business will make use of "the domestic credit system", which enables small- and medium-sized businesses that receive energy conservation support from large enterprises to sell quantities of CO_2 emission reductions.

The domestic credit system started as a governmental measure in October 2008 to be used for authenticating reductions in greenhouse gas emissions carried out by small- and medium-sized businesses *via* the provision of technology and funds from large enterprises, and for achieving goals of self-regulatory action plans and trial emission-deal schemes. Provisions of technology and funds from large enterprises to small- and medium-sized businesses are prescribed under the goals of the Kyoto Protocol Plan (sanctioned by the cabinet in March 28, 2008).

The *Ishizaka Group* switched their 20-ton class hydraulic shovel-used when dumping accumulated concrete waste into a separator — from one with a diesel engine drive to Hitachi Construction Machinery's electrically-driven shovel, the *ZX200-3*.

As a result, the company was able to reduce CO_2 emissions from what used to be approximately 137 tons a year with conventional engines to approximately 50 tons a year, achieving a drastic ratio to convention reduction of approximately 64%.

Further effects of *Hitachi*'s machinery are expected to be seen in the form of reductions in running costs, such as fuel costs and engine maintenance costs, in addition to improving workplace environments through measures such as achieving zero emissions within factory spaces and reducing the degree of heat from emissions.

The Business Model Behind the CO_2 Emissions Reduction Enterprise

As shown in Fig. 26.1, this business model will be implemented by the *Ishizaka Group*, an emission reduction credit company, in collaboration with its joint implementer, *Hitachi Capital Corporation* (Minato-ku, Tokyo, President Kazuya Miura), and associated company, *Hitachi Construction Machinery*.

In this scheme, the associated company, *Hitachi Construction Machinery* will sell and deliver an electric hydraulic shovel to the emission reduction credit company, *Ishizaka Group*, and obtain the money charged for this service. The plan furthermore calls for the emission reduction credit company, *Ishizaka Group*, to sell to *Hitachi Capital Corporation*, as domestic credits, the CO_2 emissions reduced by using the electric hydraulic shovel. Subsequently, the joint implementer, *Hitachi Capital Corporation* is to apply to the domestic credit authentication board for receiving certification that officially recognizes its emission reduction enterprise.

Hitachi Construction Machinery and the *Ishizaka Group*, upon introducing this electric shovel, filed an application to the domestic credit authentication board to receive official validation for their "CO_2 emission reduction methodology" in December 2012 and received formal certification from this board on March 23, 2011.

Future Development of the CO_2 Reduction Business

Ever since developing the electric hydraulic shovel for the first time in 1971, Hitachi Construction Machinery has developed more than 14

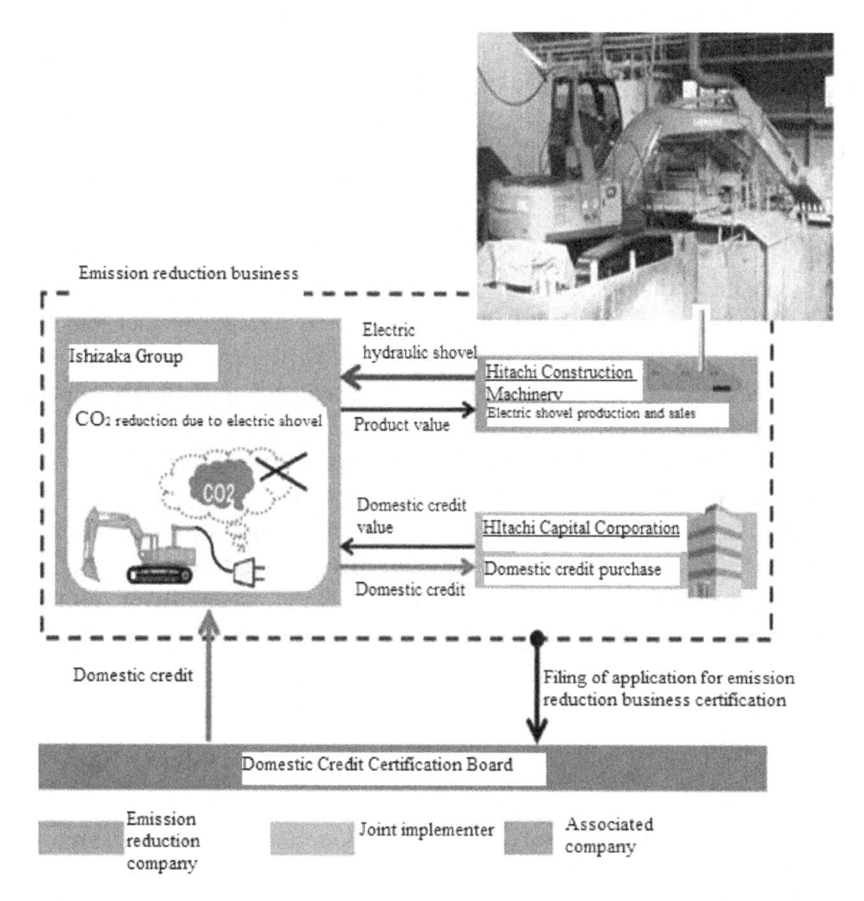

Fig. 26.1 CO_2 emission reduction business scheme utilizing the electric hydraulic shovel (Hitachi Construction Machinery, Ishizaka Group, Hitachi Capital Corporation)
Source: Hitachi Construction Machinery press release.

construction equipment models to date. The company's hard (product) strategy has entailed research and development, based on the company's original technology, for products with a high degree of value-added quality that responded to local needs. Consequently, the company has been delivering machines with superior reliability, durability and the capabilities to handle macro-scale activities at demanding sites, such as those found in the markets of emerging nations and the mining industry.

Meanwhile, for developed countries, the company has been delivering high value-added products with superior emission control for promoting a sustainable environment and superior fuel consumption and safe performance, adaptive to various job sites, such as those requiring disassembly and recycling for buildings.

As for the firm's soft strategy, the company is moving ahead with the construction of a business model that is making full use of information and communications technologies (ICT), including global e-services. The strategy also calls for the reinforcement of the support foundation for entire life cycles of products through inspections, repairs/preventive maintenance services, sales of parts and reconditioned parts, sales of secondhand cars, rentals, and financing. This is due to the fact that the cumulative number of operational machines in the market is seeing a yearly increase, raising expectations for an expansion in the sales of parts and secondhand cars.

In this way, the company has been developing businesses customized for various regions.

The system of certification using construction equipment this time is the first of its kind in the world and with this certification as a lever, the company is pursuing a strategy that will see, at the outset, the creation of a structure that will enable the provision of similar purchase incentives for all electric shovels it plans to sell in Japan.

Bibliography

1. Ishizaka-Group website. Available at http://ishizaka-group.co.jp/ (20110508)
2. IID, Inc./Response, CO_2 *reduction business using electric hydraulic shovels — itachi Construction Machinery is part of a collaboration for a globally unprecedented project*, Response IID, April 6, 2011. Available at http://response.jp/article/2011/04/06/154423.html (20110508)
3. Nikkan Kogyo Shimbun, *Hitachi Construction Machinery and Ishizaka Group realize CO_2 reductions with electric shovels and use domestic credit*, April 6, 2011. Available at http://www.nikkan.co.jp/news/nkx0120110406baai.html (20110508)
4. Hitachi Capital http://www.hitachi-capital.co.jp/ (20110508)
5. Hitachi Construction Machinery website, http://www.hitachi-kenki.co.jp (20110508)
6. Hitachi Construction Machinery press release, *Approval granted for emission reduction business using electric hydraulic shovels*, Hitachi Construction Machi-nery, April 6, 2011. http://www.hitachi-kenki.co.jp/news/press/PR20110406 10- 1124698.html (20110508)

27

CO_2 Reduction *via* the Realization of Next Generation Thermal Power

Toru Fujii

Drastic Slash in the Rate of CO_2 Emission *via* the Introduction of Advanced Ultra-Supercritical System (700°C) and Carbon Dioxide Capture and Storage (CCS)

With the global surge in energy demand conflicting with the need to carry out global warming counter-measures, an abatement in CO_2 emission rates from thermal power generation is required.

Japanese manufacturers of general electric equipment have a highly efficient coal fired power generation technology, which boasts the highest standard of its kind in the world. In such a situation, aiming to produce a next-generation thermal power generation with a zero emission rate, *Toshiba Corporation* (Minato-ku, Tokyo, President Norio Sasaki) has devised a policy called "The 2050 Toshiba Group Environmental Vision" to express the company's intention to develop technologies that can realize thermal power generation with a high level of efficiency and also capture and store CO_2 gases.

In the field of thermal power systems, as a technology for reducing CO_2 emissions, *Toshiba* is pushing forward with the development of the Advanced Ultra-Supercritical System (A-USC), which raises steam temperature from the conventional 600° level to the 700° level, while

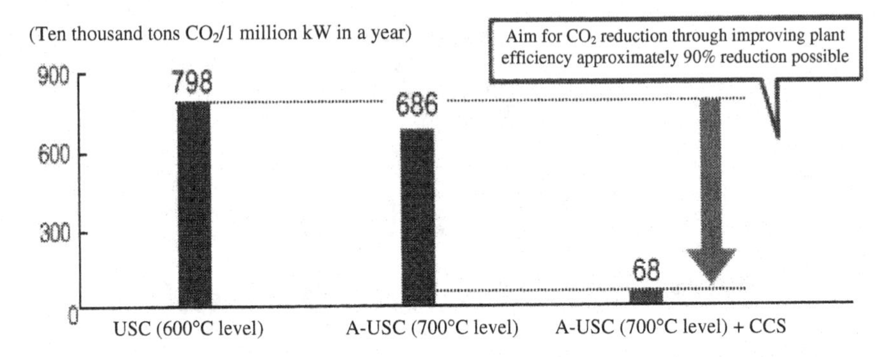

Fig. 27.1 Emission reduction quantities achieved through A-USC (700°C)+ CCS
Source: Toshiba's website.

advancing the construction of a pilot plant for the practical use of CCS technology (a power plant that captures and stores CO_2).

In this way, as shown in Fig. 27.1, by combining the A-USC system with CCS technology, CO_2 emissions from thermal power stations can be dramatically reduced to approximately 90% of conventional emissions.

Development of Next-Generation Zero Emission Thermal Power Technology

The percentage of worldwide thermal power generation by burning fossil fuel accounts for approximately more than 65% of all electric power generation, including those that use other energy sources. However, thermal power generation is responsible for the most CO_2 emissions per electrical energy unit, so the reduction of CO_2 emissions from thermal power generation is considered to be paramount.

As a measure for realizing such an environmental improvement, *Toshiba* is developing its next-generation zero emission thermal power generation technology based on the following three policies:

1. Reduction of fuel consumed through improving plant efficiency;
2. Capture and storage of exhausted CO_2;
3. Transition to fuels with minimal CO_2 emissions, such as biomass fuels.

In addition, in the development of the A-USC technology, a net thermal efficiency of more than 46% is being aimed for in tandem with a plant efficiency of 46% to 48%.

The Carbon Dioxide Capture and Storage (CCS) technology prevents the increase in the release of CO_2 into the atmosphere by isolating and capturing CO_2 emissions made by burning fossil fuel, and storing them underground or underwater.

The CCS technology enables three methods of isolating and capturing CO_2. They are as follows:

1. The combustion method that isolates and captures CO_2 from exhaust gas emissions discharged after burning with boilers or gas turbines;
2. The pre-recovery combustion system that isolates and captures CO_2 in the coal gasification process using the Integrated Coal Gasification Combined Cycle (IGCC) technology;
3. The oxygen burning boiler method that uses oxygen instead of air to make flue gases mostly CO_2. In so doing, this technology does away the need to isolate.

Toshiba's Stance on CO_2

On May 25, 2011, *Toshiba* announced that it will discuss potential development of CCS technology with the *Huadian Energy Company*, a manufacturer of equipment and machinery for power plants and a subsidiary of *China Huadian Corporation*, which is one of five major power producers in China. The joint development will be made with a view to commercialize the technology that can isolate and capture CO_2 emissions from thermal power plants.

Based on their agreement, the companies will be discussing the technical compatibility of CCS in Liquefied Natural Gas (LNG) thermal power plants, along with evaluations on its economical efficiency and concrete steps to implement the construction of a plant demonstrating the use of CCS technology in a thermal power plant.

Thermal power accounts for approximately two-thirds of all the world's power generation capacity, and it is believed to play an important role in the steady supply of electric power for the future. In addition, if we heed the warnings on safety issues that arose in the wake of the Fukushima Daiichi nuclear disaster, anticipation for thermal power plants will grow bigger in the future.

In particular, in China, the use of thermal power is extremely high, accounting for approximately 78% of all power generation in the country, making the construction of thermal power plants that respond to the challenges of global warming important for the future.

Fig. 27.2 Toshiba's pilot plant for testing its CO_2 capture technology

Source: *Toshiba*'s website.

Since 2006, *Toshiba* has been carrying out fundamental research for realizing the application of CCS technology in thermal power plants, and as shown in Fig. 27.2, the company has been carrying out since September 2009 demonstrations in the CCS pilot plant built within the Sigma Power Ariake Mikawa Power Plant (Omuta-shi, Fukuoka).

Furthermore, the *Huadian Energy Company* is also engaged in the development of CCS technology, going ahead with the construction of a CCS demonstration plant within a coal thermal power plant in the Xinjiang Uyghur Autonomous Region.

Toshiba is now working on the total optimization of thermal power plants equipped with CCS technology and is accelerating its technological development to realize its full-scale introduction by around 2020. To pursue higher efficiencies and early practical applications, both the public and private sectors need to cooperate and push forward with the development and demonstration of the technology.

Bibliography

1. IID Response, *Toshiba works toward developing CO_2 capture and storage technology for thermal power plants,* Response Seed, May 25, 2011. Available at http://response.jp/article/2011/05/25/156905.html (20110526)
2. Toshiba — Societal and Environmental activities (CSR), *Contributing toward the control of global CO_2 emissions — Toward achieving Environmental Vision 2050,* Toshiba. Available at http://www.toshiba.co.jp/csr/jp/highlight/2009/env_01.htm (20110526)
3. Toshiba Power Systems Company, Thermal/Hydraulic Power Division, *CCS (CO$_2$ capture and storage technology — The technologies we're developing,* Toshiba. Available athttp://www.toshiba.co.jp/thermal-hydro/thermal/approach/ccs/index_j.htm (20110526)
4. Kiyoshi Miyaike, *Next-generation Thermal Power Generation Technology and Future Trends in the Supply of Electric Power Energy,* The Toshiba Review, Toshiba, Vol. 63, No. 9, 2008, pp. 2–7.

28

Introduction of the Electric Bus Infrastructure

Toru Fujii

Development of the Replaceable-Battery Based Electric Bus Infrastructure

Mitsubishi Heavy Industries, Ltd. (Minato-ku, Tokyo, President Hideaki Omiya) succeeded in the development of an electric bus infrastructure powered by replaceable batteries, and its demonstration is to be held in Kyoto-shi and Aomori, where the Ministry of Land, Infrastructure and Transport is carrying out trials of environmentally sustainable cities.

For Japan, as a policy to combat global warming, arriving at measures that help realize the drastic reduction of CO_2 emissions is an urgent matter from now on. In particular, measures for reducing CO_2 emissions generated from automobile-related businesses, which account for approximately 20% of all CO_2 emissions, are deemed important, and the company is proposing to carry out municipal measures led by local governments for raising the level of convenience through securing the regularity and expediency of mass-transit electric buses, and also improve the level of convenience by installing in-vehicle information communication inside zero-emission, low-noise, highly comfortable electric buses.

Mitsubishi Heavy Industries aims to reduce CO_2 emissions of the transport sector with the activation of mass transit and reduce the number of people who have no means of transportation. Specifically, the company plans to do so by operating electric buses to promote modal shifts from driving private automobiles to making use of public transportation. By increasing the number of bus users in this way, it will also be possible to improve the bottom line of the bus transit business, which is currently in the red. In addition, by adopting an electricity storage system that makes use of replaceable batteries for electric buses, the company will be able to promote the use of a renewable energy source and contribute toward the realization of a low-carbon footprint society.

CO_2 Footprint Per Passenger Dramatically Reduced to One-Eight

As shown in Fig. 28.1, the merit of converting the energy of mass transit buses lies in the fact that approximately 30 tons of CO_2 a year could be reduced per bus. That in itself will have a large effect on the overall reduction of CO_2 emissions.

In addition, by having people switch from driving private automobiles to riding on electric buses, CO_2 footprint per passenger can be reduced to around one-eighth the conventional rate. For this reason, the effect of the modal shift caused by introducing the electric bus is great.

For this reason, to achieve the goal of CO_2 reduction, it is necessary to not only turn the bus into an electric vehicle, but to actively carry out a modal shift from driving private cars to riding mass-transit buses.

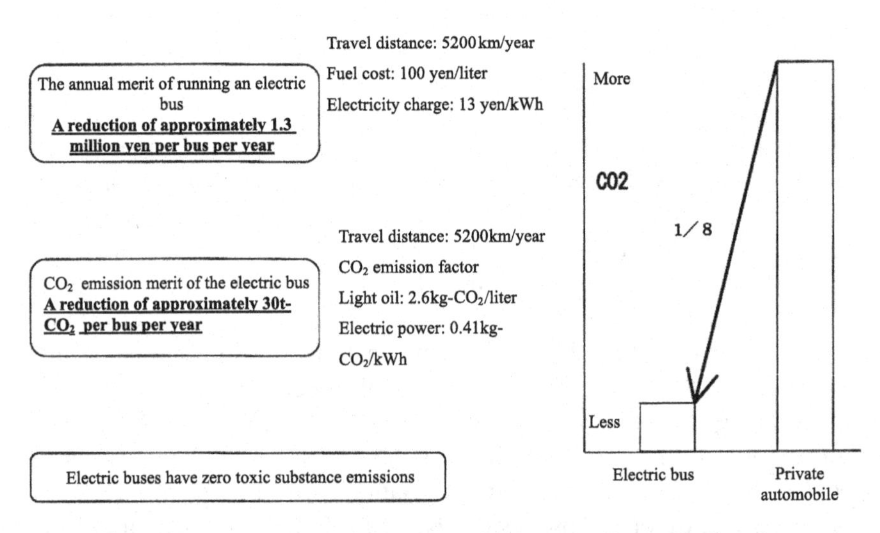

Fig. 28.1 Merits of the electric bus and the per-passenger CO_2 footprint
Source: Extracted from "The Electric Bus Infrastructure Project" by Yoshitaka Kakuhama *et al.*

To this end, the key is to attempt the diffusion of the electric bus. To do so, it is necessary to activate public bus transportation by improving the convenience and comfort of shuttle buses and by carrying out forms of town planning that account for guiding passengers from private automobiles to public transportation.

In Japan, approximately 80% of public bus operations across the nation is said to be in the red, making the business environment for shuttle bus operators a harsh one. The main cause of this is surmised to be attributable to the rise in the number of private automobiles in urban areas. In effect, this rise is contributing to traffic jams that make it difficult to ensure that shuttle buses arrive at scheduled times. For this reason, passengers choose to avoid riding on buses, viewing the shuttle bus to be an inexpedient mode of transportation.

While from now on we can presume that the efficient use of electric power will be increasingly promoted, by allowing each area to build the infrastructure for electric buses, it will become possible not only to raise the total energy efficiency rate, but also to dramatically bring down the rate of CO_2 emissions to one-eighth its conventional rate. Specifically, the adoption of electric buses can help decrease the number of existing bus services, integrate bus routes and thereby reduce the degree of dependence on private automobiles from the level of one person per car.

According to a specific proposal made by *Mitsubishi Heavy Industries*, as shown in Fig. 28.2, by installing exclusive bus lanes and signals that give priority to public transport, the company intends to regulate entry of non-electric vehicles into the city, in addition to providing park-and-ride facilities in and around urban areas.

In the area of the in-vehicle database service for electric buses, the company plans to make use of the information and communication technology of Intelligent Transport Systems (ITS) to offer real-time information useful when boarding buses, such as arrival times, transit times and traffic information, in addition to other information sought by passengers, such as, in the case of tourist destinations, the weather, map searches and sightseeing information.

Challenges of Introducing the Electric Bus

What is indispensable in the running of electric buses is the installation of charging facilities for charging lithium ion batteries. While the mainstream charging method for battery-installed electric cars is the

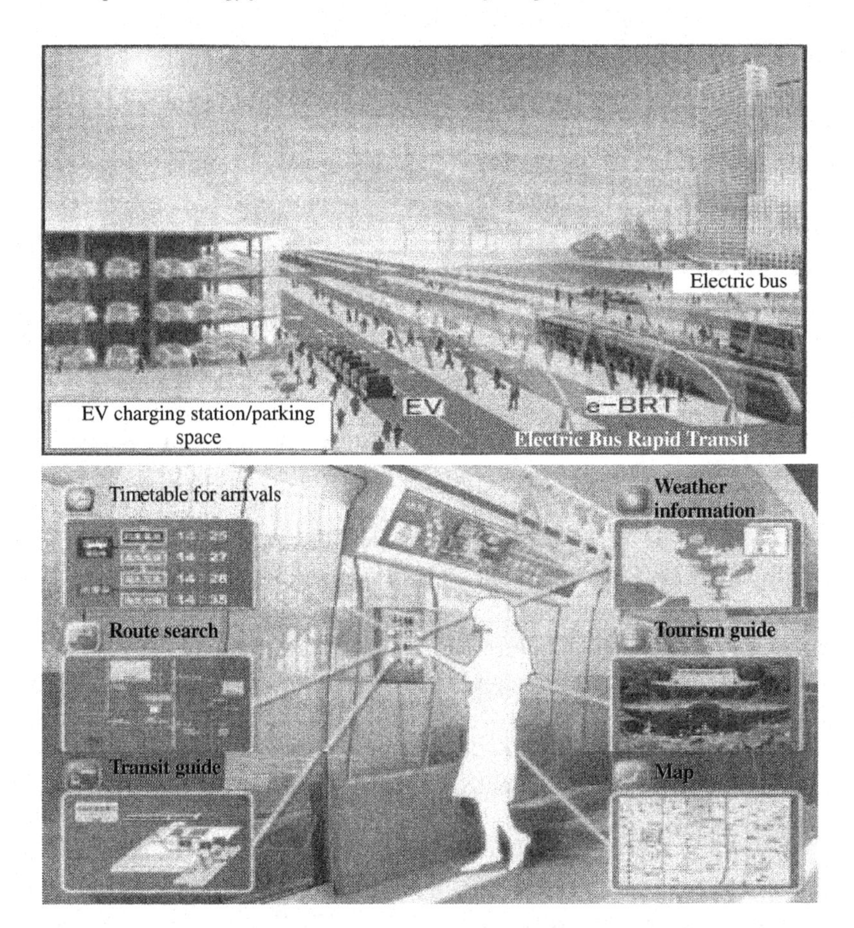

Fig. 28.2 Park Island and on-board database service for electric buses

Source: Yoshitaka Kakuhama, "The Electric Bus Infrastructure Project".

quick-charge plug-in method, adopting this for electric buses poses the following two challenges.

1. Recharging takes anywhere from dozens of minutes to one hour, putting the bus out of service in the meantime. Additionally, a parking space will become necessary for the purpose of recharging.
2. Recharging facilities need to be augmented since the per bus demand for power is irregular and large in the daytime, amounting to about 100kW (kilowatts). For this reason, the amount of power supply agreed with electric power companies needs to be reviewed.

To respond to these challenges, *Mitsubishi Heavy Industrie*s is proposing to adopt a design that will allow the bus' battery to be removable and therefore rechargeable externally. In essence, the battery will be replaceable.

The Future of the Electric Bus Infrastructure

With Japan's adoption of a policy for fostering environmentally sustainable innovation (or what is known as the green-innovation strategy), the implementation of various measures aimed at realizing drastic reductions in CO_2 emissions is said to be a pressing need. To this end, realizing a modal shift from car driving to bus riding will prove significant.

Mitsubishi Heavy Industries is preparing to launch the electric bus service by carrying out trial runs in Kyoto and Aomori. Specifically, they will have electric buses actually servicing routes in those areas, while measuring their CO_2 reduction effects, issuing questionnaires to passenger monitors, and carrying out investigations into the impact of the buses on the surrounding traffic and pedestrians.

In the future, where new urban developments such as smart cities and smart communities are concerned, it can be said that the role played by the electric bus infrastructure will be great.

Bibliography

1. Yoshitaka Kakuhama, Jin Kato, Yasushi Fukuizumi, Masaharu Watabe, Takashi Fujinaga, Takuya Tada, *Special Feature on New Products/New Technologies — The Next Generation Public Transportation that Contributes to Low Carbon Society-The Electric Bus Infrastructure System Project-,* Mitsubishi Heavy Industries Technical Review, Mitsubishi Heavy Industries, Vol. 48, No. 1, 2011, pp. 2–5. Available at http://www.mhi.co.jp/technology/review/pdf/481/481002.pdf (20110508)
2. Mitsubishi Heavy Industries website. Available athttp://www.mhi.co.jp/index.html (20110508)

29

Twitter-Driven Record-Breaking Profit

Junpei Nakagawa

Switching to a Strategy of Establishing Urban Stores

Yamada Denki, Co. Ltd. (Takasaki-shi, Gunma, President Tadao Ichinomiya), based in Gunma, developed as a suburban mass retailer chain of consumer electronics running roadside stores. It prevailed in the intense price competition with *Kojima, Co., Ltd.* (Utsunomiya-shi, Tochigi, President Etsuo Terasaki) located in the neighboring prefecture, and has become the largest consumer electronics mass retailer in Japan today. Amid an environment that saw all industries across the board being heavily damaged by the Great East Japan Earthquake, the company's financial results ending March 2011 showed a 68.5% gain in the company's net profit compared to the previous year, breaking its previous record for the highest profit earned by the company.

One of the factors behind this gain is the company's foray into in the urban store sector and its establishment there. With suburban stores, the company was able to have an edge over the competition by realizing a scale merit through devising a cost-leadership strategy that focused on heavy buying and mass selling of major appliances. On the other hand, for consumer electronics products, since consumers are highly discerning about them, the know-how of urban mass retailer camera chains was used. Such chains included *Yodobashi Camera Co. Ltd.* (Shinjuku-ku, Tokyo, President Akikazu Fujisawa), which handles a large selection of products.

Yamada Denki launched its urban store *LABI Namba* in Osaka in 2006. This is because the company was compelled to compete against new competitors that differed from their conventional ones, so a new competitive strategy was needed. Such new competitors included *Joshin Denki, Co., Ltd.* (Naniwa-ku, Osaka-shi, President Katsuhiko Nakajima)

Yamada Denki has achieved its growth by carrying out drastic price cuts for a large volume of goods in stock, using the catchphrase "cheaper than other stores". However, with consumer electronics goods, the key to gaining a competitive edge lies in how fast the company can stock and offer new goods rather than in carrying out price cuts. Consequently, the company set its eyes on the use of *Twitter* to deliver information on the arrival of new shipments and went on to set up accounts for all of its *LABI* stores.

Store-Specific *Tweets* on the Receipt of New Shipments of Goods in Short Supply

As a successful instance of the appropriate tweeting of the status of product inventories within Japan, there is the case of the second-hand store of *Fujiya Camera* (Nakano-ku, Tokyo, President Kohjiro Ohtsuki), whose tweeting led to a 20% increase in its mail-order sales. The camera is a luxury item and if it is second-hand, its condition will vary even among units of the same model. For this reason, the camera is a one-off item. Since providing real-time information for the user would mean that his or her opportunities for purchasing a desired product would rise, he or she wouldn't have to diligently visit stores to keep checking for its availability. This resourcefulness was largely given coverage by a TV program on TV Tokyo Channel 12.*

Even *Yamada Denki* began to tweet on the supply situation for consumer electronics products sold in its urban *LABI* chain of stores. In particular, the company is actively providing information on popular, new products on a store-specific basis, having set up *Twitter* accounts for each of its *LABI* outlets. While there are often cases where only shipments of small quantities of popular new products — such as *Apple's iPhone* and the *Nintendo DS* — arrive at stores, prolonging their shortage for a long time, by receiving appropriate *tweets* on the supply status of products, customers can acquire real-time information such as information that indicates which store a certain product is available at. For this reason, enthusiastic buyers of new products can purchase them easily without having to visit many stores to search for them.

Such buyers comprise the "innovator/early adopter" group and a great deal of them have blogs and *Twitter* accounts. And in many cases, when they are able to make early purchases of popular items, they tend to post messages such as "I was able to buy the product, thanks to *Yamada Denki*'s *tweet* on its supply status".**

As a result, users accessing the websites of leading reviewers — the people who evaluate products and are therefore said to hold the key to their spread and popularization — become followers of *Yamada Denki*'s tweets and wait for product supply information from the company's stores. And as shown in Fig. 29.1, after they receive *tweets* pertaining to the supply situation of a product from any of *Yamada Denki*'s stores, the likelihood of their making a purchase rises significantly. This is a synergistic effect.

As can be seen above, *Yamada Denki*'s strategy mainly revolved around receiving high-volume shipments and making mass sales. For such a company to win the trust of customers of urban stores where the type of popular items are different from those they were accustomed to selling, it was necessary to increase the opportunities for making purchases at *Yamada Denki* by conveying information on the supply situation of products, even if such information pertained to shipments of only a small supply. At present, it has become common to see posts on word-of-mouth sites such as *Kakaku.com* (Price.com), indicating that someone was able to buy a particular product at *Yamada Denki* when it was on backorder at other stores.[1]

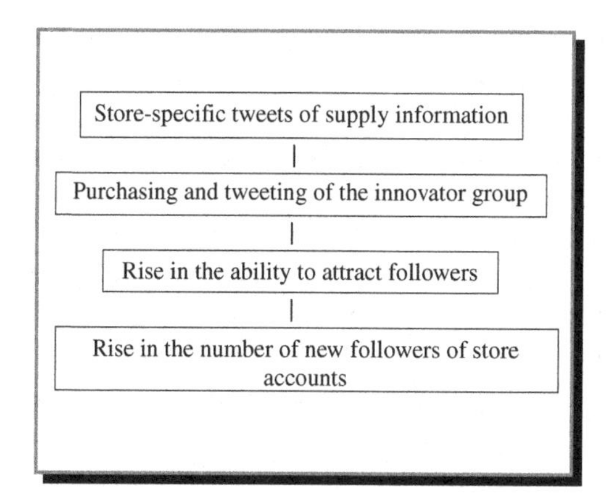

Fig. 29.1 *Yamada Denki's Twitter* effect

Source: Author.

A major factor behind the company's successful establishment in the urban store sector is the company's decision to use its first-rate nationwide chain of stores and provide real-time information to its *Twitter* followers regarding the availability of popular products at their stores.

Using *Twitter* to Announce Special Limited-Time Sales not Mentioned in Flyers

By successively tweeting store-specific, special time sale information, *Yamada Denki* attracted follower traffic in the vicinity of their stores, realized an accurate information service, and heightened the marketing effect. The competition between *Yamada Denki* and *Kojima* saw both companies featuring loss leaders in their advertising literature and providing information on deep discounts for a limited number of people during a limited period of time. Consequently, the two companies vied for the most business on weekends when lines of customers would form prior to the opening of stores. However, customer traffic tended to gradually decline from early afternoon — when loss leader items were sold out — through the evening. On the other hand, on weekdays, compared to the time of night when people were returning home, daytime customer traffic tended to be little. In addition, when it rained, the pace of customer traffic slowed down further.

Consequently, by announcing *via Twitter* special time sales taking place mainly during the evening, *Yamada Denki* is now attempting to level the number of customer visits. This is a guerrilla marketing approach aimed at price-sensitive users, which the company is intrinsically strong at carrying out. In effect, real time announcements are being made *via Twitter* in sync with favorable weather conditions and customer numbers. Followers receiving tweets from stores in their vicinity can purchase products at bargain prices without standing in lines and stores can make their sales grow during time zones when sales amounts tend to be low.

Lucky users who are able to purchase through special time sales post messages on *Twitter* or word-of-mouth sites, informing that they were able to buy because they had received special information *via Twitter.*

And as shown in Fig. 29.2, since other users who read these messages become new followers of *Yamada Denki*'s *Twitter* accounts, a synergistic effect is produced and the special time sales is given a further boost. Even *Yamada Denki*, a company that grew to become Japan's top suburban

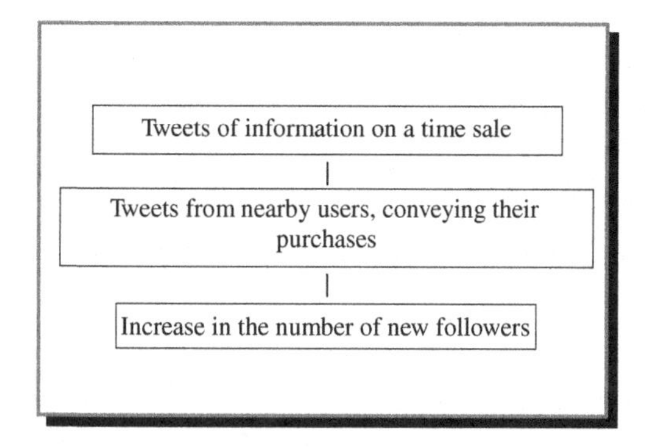

Fig. 29.2 The synergistic effect of the time sale

Source: Author.

consumer electronics mass retailer chain by successively carrying out expansion strategies, has prioritized the cultivation of new customers to achieve further development.

While the use of *Twitter* in the *LABI* chain of urban stores is showing promising signs for realizing greater possibilities, there are concerns that profits will dip in the wake of the termination of the consumer electronics eco-point system. Therefore, if there can be an increase in the number of opportunities for having direct information exchanges with followers — opportunities such as after sale services — it will become possible to build even further trust among customers.

Notes

*"World Business Satellite," broadcasted on June 28, 2009.

**The innovator group refers to the group of consumers who purchase new products of their own accord and without fearing any risk associated with their purchase. In effect, they are the buyers of a product life cycle's introductory period. Early adopters refer to the group of consumers who adopt the new products after the innovators, purchasing them after deliberation.

Bibliography

1. Kakaku.com website. Available at http://kakaku.com/ (20110610)

30

The Globalization of Inter-Regional Mobility Through Low-Cost Carriers

Kazuo Matsude

The Super Discount Airlines of Europe

When I was stationed in London from 2004 through 2009, I frequently used low-cost carriers out of personal interest. At the time, EU's airline liberalization was underway and price reductions of fares were remarkable too. Amid such an environment, the one low cost carrier (LCC) that stood out among others in terms of its price tag was the Irish airline, *Ryanair*.

Ryanair's ticket prices are often lower than the prices of other companies and *Ryanair*, in particular, offers low-priced fares during limited-time sales. For example, this author investigated *Ryanair*'s website on the early morning of April 17, 2011 and found out that if one made a reservation by April 18 for a flight departing from Stansted airport in London for Rygge airport in Oslo on May 5, he or she would only be charged the airfare of 7 pounds (945 yen when 1 pound = 135 yen). This price included miscellaneous expenses. Meanwhile, a search on the Internet travel site *Opodo* for other Oslo-bound flights departing from London on May 5 showed the cheapest flight to be *Norway Airline*'s 46 pounds (6,210 yen), followed by *Scandinavian Airlines*' 69.5 pounds (9,383 yen), and *British Airways*' 71.8 pounds (9,693 yen).

Ranking	Airlines (country)	Passenger number
1	Southwest Airlines (U.S.)	101,338
2	American Airlines (U.S.)	85,720
3	Delta Air Lines (U.S.)	67,95
4	China Southern Airline Group (China)	65,959
5	Ryanair (Ireland)	65,282
6	United Airlines (U.S.)	56,024
7	Lufthansa (Germany)	53,223
8	US Airways (U.S.)	50,975
9	Air France (France)	47,965
10	Continental Airlines (U.S.)	44,032

Fig. 30.1 Ranking of airlines by the number of passengers (2009) (Unit: 1,000)
Source: IATA, *The World Air Transport Statistics 54th edition.*

While the flagship carriers of various countries are on the bandwagon of the bargain-fare trend too, *Ryanair*'s price discounts are certainly extreme.

In effect, *Ryanair* is increasing its repeat passengers by frequently carrying out bargain sales mainly on weekends. As a result, as shown in Fig. 30.1, according to the number of travelers by airline, *Ryanair* ranks at the top, and is fifth in the world even. Incidentally, the world's No. 1 airline is the US's LCC, *Southwest Airlines.*

Ryanair's pricing stands out because, unlike other airlines, it has adopted a management strategy that concentrates on a low-price strategy in particular. For example, the company has standardized all its aircrafts to the latest medium-sized Boeing 737–800 models, while streamlining their maintenance and education systems. Such measures have not been adopted by any other major airlines. In addition, a well-known cost-reduction measure taken by *Ryanair* is its installation of non-reclining seats with no seat pockets. Incidentally, safety cards are stuck on the back of the seats. Such measures aim to realize not only equipment cost reductions, but also cleaning-time reductions.

Furthermore, as *Ryanair*'s key strategy, there is the company's use of secondary airports — airports that are located in the outskirts of large cities that have cheap airport and landing taxes — instead of airports located in city centers. This was exactly the case with the London-Oslo flight mentioned earlier. Both London Stansted Airport and the Rygge Airport (located in Oslo) are located at places that take more than one hour to reach *via* bus. Consequently, *Ryanair*'s share at such airports became larger and helped to strengthen the company's clout to bargain with them.

For example, during the enforcement of a bargain sale, while cutting down the airfare, *Ryanair* tends to also seek reductions in the landing taxes imposed by the airport. In the above-mentioned example of the London-Oslo flight, the amount that taxes and commissions comprised in the fare of seven pounds was zero. Based on this evidence, when *Ryanair* carried out its bargain sale in that particular case, we can see that the airport had helped *Ryanair*'s business by exempting the company from paying charges such as the landing tax.

Low-Cost Airlines in Japan

Low-cost airlines in Japan whose ticket prices are below those of the regular fares of the two major domestic airlines, *Japan Air Lines Co., Ltd.* (Shinagawa-ku, Tokyo, President Masaru Onishi) and *All Nippon Airways Co., Ltd.* (Minato-ku, Tokyo, President Shinichiro Ito) include *Hokkaido International Airlines Co., Ltd.* (Chuo-ku, Sapporo, President Sadao Saito), *Skymark Airlines Inc.* (Ota-ku, Tokyo, President Shinichi Nishikubo), *Skynet Asia Airways Co., Ltd.* (Miyazaki-shi, President Yutaka Morinaga) and *STARFLYER Inc.* (Kitakyushu-shi, Fukuoka, President Shinichi Yonehara). However, there are no airlines to be seen with a pricing scheme as extreme as *Ryanair*'s scheme. Additionally, even the two major airlines, due to the early reservation discount system, have been extensively offering airfares that are cheaper than normal airfares. For this reason, the actual price differences between the airlines are not that great.

Meanwhile, when we look at international airlines, we see that they are rather actively flying to Kansai Airport and other regional airports, as shown in Fig. 30.2. Among Japanese airlines, *All Nippon Airways* is finally at the stage where it will be launching its own low-cost carrier next spring by partnering with investment companies in Hong Kong.

Among foreign carriers, Malaysia's *Air Asia* caused a stir when it launched its Haneda to Kuala Lumpur service and offered a 5,000 yen one-way airfare. However as of April`17, 2011 this airfare was 16,000 yen, which is around half the price for discounted airline tickets of major carriers such as *JAL*. While it is undoubtedly cheap, it doesn't reach extreme bargain levels. Along with the point of choosing to land at Haneda Airport for its convenience, there appears to be a distinct line drawn between this

Airport name	Airline name (country)
Narita	Air Asia (Malaysia), Jetstar (Australia)
Ibaraki	Spring Airlines (China)
Kansai	Jetstar (Australia), Jeju Air, AIR BUSAN (Korea) Jetstar Asia Airways (Singapore)
Fukuoka	AIR BUSAN (Korea)
Kitakyushu	Jeju Air (Korea)

Fig. 30.2 List of low-cost carriers by the airport they fly into

Source: Original research by this author.

airline's business model and *Ryanair*'s business model, which thoroughly pursues a low cost strategy.

It is rather in the models adopted by *Spring Airlines* and *Jeju Air* that we can see the partial adoption of *Ryanair*'s business model. Both of these airlines have selected the secondary airports, Ibaraki Airport and Kitakyushu Airport.

The Future of the Budget Pricing Oriented LCC Business

In conclusion, full-scale budget oriented low cost carriers such as *Ryanair* have not yet developed in Japan, making them businesses that have room to grow in the future. In their business model, the implementation of a low-price strategy that realizes overwhelmingly low airfares is necessary and the selection of secondary airports is indispensable. As candidate sites for such airports in Japan, it includes Ibaraki Airport located in the capital territory, Kobe Airport located in the Kansai area, and Saga and Kumamoto located in Kyushu.

Each airport and the local authorities of the airports should offer incentives such as reduction in landing fees to *Spring Airlines* and *Jeju Air* to attract them. These airlines are considered to be foreign airlines that have adopted strategies close to *Ryanair*'s strategy. In fact, the Governor of Kumamoto, Ikuo Kabashima has paid a visit to *Spring Airlines* last autumn and requested the launch of flights servicing the Kumamoto-Shanghai route, attempting to tie them to regional vitalization.

At the same time, the prospects are very good that a Japanese airline with a super low-cost structure such as *Ryanair*'s structure will emerge, grow, and successfully make a foray into servicing international routes by establishing hubs in secondary airports. For example, airports such as the Ibaraki Airport and the Kumamoto Airport should be turned into hubs for airlines such as *Skynet Asia Airways* and *STARFLYER*.

With that in mind, the Japanese low cost carrier could enter into business and capital tie-ups with low cost carriers based in Korea, China and Southeastern Asian countries and jointly implement medium and small-sized flights into each other's territories. In doing so, the Japanese low cost carrier could very likely realize a strategy that can help boost the turnover rate of its aircraft fleet and secure profits.

Bibliography

1. IATA, *The World Air Transport Statistics 54th edition,* 2010.
2. Asahi Shimbun, *Budget carriers affiliated with All Nippon Airways will fly from Kansai to Fukuoka/New Chitose/Seoul,* Asahi Shimbun Electronic Edition, April 13, 2011.
3. The Tourism Exchange and International Affairs Division of Kumamoto Prefecture, "A delegation led by the governor has visited China," The press archives of the Kumamoto Prefectural Office, Kumamoto prefecture, November 4, 2010.
4. Takuya Sumikawa, *Skymark takes steps to enter into the international airlines sector by 2002 with the A380,* Asahi Shimbun Electronic Edition, November 9, 2010.

31

Achieving a 15% Savings in Electricity without Reducing Factory Output

Tetsuro Saisho

Power Saving Measures in the Aftermath of Major Earthquakes

Among power saving measures put into effect due to the impact of the 2011 Tohoku earthquake and tsunami disaster, there has been a case in which power conservation was implemented without reducing industrial output. In particular, companies in the service areas of Tokyo Electric and Tohoku Electric Power uniformly received the advice to seek a 15% reduction in power consumption. Consequently, these companies went on to attempt such a cut by avoiding what they deemed to be wasteful uses based on in-house electricity usage data.

In the aftermath of the Tohoku earthquake and tsunami disaster that occurred on March 11, 2011, the Government authorities on May 13, 2011 officially recommended all of the companies in the service areas of *Tokyo Electric Co., Ltd.* (Chiyoda-ku, Tokyo, President Masataka Shimizu) and *Tohoku Electric Power Co., Inc.* (Sendai-shi, Miyagi, President Makoto Kaiwa) to uniformly reduce their power consumption by 15%. It was to be adopted as a measure to deal with the summer's electricity shortage. While the government relies on large corporations to take voluntary steps to this end, they are preparing to enforce usage restrictions based on Article 27 of the Electric Utility Industry Law. In response to the request from the Government to save power, various conservation measures have become full fledged in the industrial world.

In the case of *Panasonic Co., Ltd.* (Kadomashi, Osaka, President Fumio Otsubo) all thirty of the company's plants serviced by Tokyo Electric and Tohoku Electric Power have achieved the standard of reducing their electricity consumption by 15% while maintaining their current production volumes. This was made possible by installing in *Panasonic*'s factories sensors that measure electricity usage to gather data for the purpose of identifying and avoiding wasteful uses of electricity.[1]

As shown in Fig. 31.1, in the industrial world there are companies other than *Panasonic* that are aiming to reassess the value of holidays and working methods, decentralize by transferring production and equipment, introduce energy-saving equipment, implement rotational holidays in which workers unanimously take two days off during the working week instead of on Saturdays and Sundays, implement night, early morning and weekend shifts when electricity usage tend to be low, and strike a balance between the need to ensure business continuance and the need to reduce power consumption by prolonging summer holidays.

Power-Saving Measures of Companies

In the reduction of electricity consumption, plenty of technologies and products related to energy conservation and electric power saving have been proposed and carried out. For example, *Microsoft Japan* (Minato-ku, Tokyo, President Yasuyuki Higuchi) has developed an automatic alteration program that helps reduce electricity consumption by 30% in personal computers running the Windows operating system. *Microsoft* is spreading this program to companies and households with the cooperation of PC manufacturers such as *Nippon Electric Co., Ltd.* (NEC) (Minato-ku, Tokyo, President Nobuhiro Endo) and *Fujitsu Ltd.* (Kawasaki-shi, Kanagawa, President Masami Yamamoto).

Although IT-related equipment have become indispensable for the modern enterprise, it is believed that they account for more than 20% of energy consumption in office buildings. Since it has been estimated that the number of personal computers running Windows (which are IT-related equipment) in *Tokyo Electric*'s service area is approximately 24,550,000, the consumption of approximately 330,000 kW of power can be reduced if *Microsoft*'s automatic alteration program were to be installed in all of those personal computers.

With regard to transferring equipment and production to respond to the need to save power in the aftermath of the 2011 Tohoku earthquake and

Reassessments of holidays and ways of working	
Toshiba Corporation	Extending holidays and fragmenting days off by every workplace (30% during peak periods), transfer to summer holidays
Sony Corporation	Transfer all holidays to summer holidays to consider introducing daylight-saving time.
KomatsuLtd. Co., Ltd.	Considering the introduction of a three-day weekly holiday schedule for employees of the company's Tokyo headquarters.
Nikon	Increase summer holidays in exchange for giving up holidays during springtime Golden Week holidays.
Sumitomo Heavy Industries	Switch to days-off/swing shifts at three factories in the capital territory.
Sumitomo Metal Industries, Ltd.	Carry out a daylight-saving schedule that will see the start and close of offices one hour earlier for the company's headquarters in both Osaka and Tokyo.
Taisho Pharmaceutical Co., Ltd.	Set up a rest day during weekdays and operate on Saturdays and Sundays.
Teijin Ltd.	Allow employees working in the capital territory to telecommute.
Terumo Corporation	Extend nighttime work shifts and operations of in-house power generating stations.
Makino Milling Machine Co., Ltd.	Bring forward operations and starting times also during Golden Week holidays to six o'clock, excluding Saturdays and Sundays.
Mitsui Chemicals, Inc. Co., Ltd.	Carry out operations in the Mobara branch factory on Saturdays and Sundays.
Mitsui Engineering & Shipbuilding Co., Ltd.	Consider shifting the Chiba establishment's welding processes to take place during nighttimes/days-off.
Showa Denko K.K.	Switch magnet alloy-related operations to nighttimes/days-off.
Ion Co., Ltd.	Consider starting working hours of stores one hour earlier for stores starting at 9:00 a.m.
Transfer of production and facilities	
Fujitsu Ltd.	Transfer 30% of development servers in the capital territory to Hyogo and Toyama.
Cherry Internet Co., Ltd.	Transfer a portion of server functions to Kansai.
NTT Data Corp. Co., Ltd.	Transfer servers in data centers (DCs) in Tokyo to DCs in Osaka and Nagoya.
Honda Motor Co., Ltd.	Change the production base of the new model car, "FIT SHUTTLE," from Saitama to make it triplex.

(*Continued*)

(Continued)

KomatsuLtd. Co.,Ltd.	Carry out full-scale operations of in-house power generating stations at two factories, and hold meetings at shutdown factories (40% level).
Nippon Steel Corporation	Consider intensive implementation of periodic maintenance operations of facilities, such as steel material rolling, during summertime.
Toray Industries, Inc.	Temporarily carry out alternative production of in-house power generating stations in foreign countries and in the western part of Japan.
JFE Steel Corporation	Promote the reinforcement of production in areas besides Tokyo Electric Power and Tohoku Electric Power's areas of jurisdiction; areas such as in the western part of Japan.
Introduction of energy-saving equipment, etc.	
JX Nippon Oil & Energy	Switch to energy conservation lamps at approximately 850 stands located in Tohoku Electric and Tokyo Electric Power's juris dictions.
Marubeni Corporation	Replace the lighting of the headquarters building with LED and switch to thermally insulated window shades with a budget of more than 200 million yen.
Seven-Eleven Japan Co., Ltd.	Introduce sensors detecting electricity usage into 6,000 stores, and LED lights to 5,000 stores.
ShochikuCo.,Ltd.	Assign rotational schedules for movie screenings. Dim down Shimbashi Enbujo's stage lighting (up to 50%).
Coca-Cola Japan Co., Ltd., DyDo DRINCO, Suntory Holdings Limited, ITO EN, Ltd., Kirin Beverage Company, Limited, and 16 companies that belong to ASAHI SOFT DRINKS CO., Ltd.	Stop the cooling of approximately 870,000 soft drink vending machines (under Tokyo Electric's jurisdiction) in rotation from five to six hours per day.

Fig. 31.1 Restrictive electric power measures of major companies

Source: Created by revising the data that appeared in the May 14, 2011 edition of the "Nihon Keizai Shimbun".

tsunami, *Toray Industries, Inc.* (Chuo-ku, Tokyo, President Akihiro Nikkaku) is considering transferring production to the western part of Japan and overseas destinations while maintaining its rate of operation in addition to reinforcing in-house power generating stations. In addition, *JFE Steel Corporation* (Chiyoda-ku, Tokyo, President Eiji Hayashida) has indicated that it will consider transferring its production base to a place outside of Tokyo Electric's and Tohoku Electric Power's service areas and reinforce its output as an iron & steel manufacturer in the western part of Japan.

A power-saving measure that is spreading among companies is the measure to reexamine various aspects beginning with the working methods and patterns of their employees. *Sumitomo Metal Industries, Ltd.* (Osaka-shi, Osaka/Chuo-ku, Tokyo, President Hiroshi Tomono) introduced daylight-saving time for moving up working hours one hour ahead; *AEON Co., Ltd.* (Chiba-shi, Chiba, President Gennya Okada) moved up the opening and closing times of stores that open at 9:00 a.m., one hour ahead for the period from the end of June to the end of September.

In addition, *Teijin Ltd.* (Osaka-shi, Osaka/Chiyoda-ku, Tokyo, President Shigeo Oyagi) is planning to cut back on its office's electricity consumption by introducing a telecommuting system for approximately 2,000 of its office staff working in the capital region.

Meanwhile the communication industry is having a hard time arriving at appropriate countermeasures since companies in this sector cannot afford to shut down their machinery and equipment. *Nippon Telegraph and Telephone Corp.* (Chiyoda-ku, Tokyo, President Satoshi Miura) and *Softbank Corporation* (Minato-ku, Tokyo, President Masayoshi Son) are attempting to curtail their electricity consumption in their offices by more than 30% through the introduction of telecommuting. At the same time, with regard to their communication equipment, they intend to demand the Government to allow them the flexible application of restrictions, which also entail them to contract their equipment out.

Panasonic's Power Saving Measures

The 2011 Tohoku earthquake and tsunami disaster triggered the tightening of the electric power supply and put pressure on companies to cut down on their power consumption by 15%. However, it has been difficult for them to grasp just how much and in what ways electricity can be saved by reducing consumption for lighting, air conditioning, and personal computer usage. Additionally, carrying out investigations for the purpose of grasping these points is also proving difficult.

Amid such a situation, *Panasonic* will carry out power conservation measures and reduce 15% of electricity consumption in all thirty of its plants while maintaining its current output levels. This measure will increase the number of holidays for the factories and enable the company to curtail power consumption by 15% without transferring production to another area.

The factories targeted for carrying out power conservation are the Mobara plant (Mobara-shi, Chiba) and the Utsunomiya plant (Utsunomiya-shi, Tochigi) where the production for the company's video products and display devices are carried out; the Sendai plant (Natori-shi, Miyagi), the Yamagata plant (Tendo-shi, Yamagata), and the Fukushima plant (Fukushima-shi), which handle production for the company's network business. Carrying out power conservation at these plants will have the merit of avoiding any stagnation in business activities. The staff officer in charge of implementing power-saving measures will visit each plant and oversee the installations of sensors and the introduction of a power-saving simulator to grasp the usage of electricity, gas and water, and gauge temperature and moisture levels by every product line and equipment.

A power saving simulator is a system that can simulate for factories, buildings, and stores simple forecasts of power saving quantities and power saving measures that could be taken during peak periods based on the data collected through the sensors. In addition, this system estimates how much power is consumed by lights, air conditioners and electric outlets by analyzing electricity and gas usage data entered in monthly invoices, making it possible to create graphs comparing the previous year's data on reductions; simulate usage of lights, air conditioners and electric outlets; and simulate power saving measures that can be taken during peak consumption periods.

This system, in addition to accommodating a proprietary software that performs data analysis and simulations of power-saving measures, computes approaches that can help each factory continue their operations in the most waste-free, efficient ways possible. In addition, for the future, the company plans to provide support for cutting back on the usage of lights, active deactivation of power sources, reassessments of temperature control for clean rooms, reductions in the number of compressors and brief shut downs.

This system itself will not guarantee reductions in power consumption but it will be prove to be an effective tool for roughly grasping how much conservation can be realized. Furthermore, if detailed data pertaining to buildings such as data on air conditioning systems could be entered, an even more detailed energy conservation simulation would become possible. With these power saving measures, *Panasonic* aims to reduce its power consumption (during peak periods) this summer by 15% less than its previous year's level of consumption.

Under the terms of contract with Tokyo Electric and Tohoku Electric Power, *Panasonic* is allocated in total 160,000 kW of electric power. In the past, the company had made a similar attempt for several of its bases, such as for the Wakayama plant (Kinokawa-shi, Wakayama), and believes it can achieve this consumption level again. The investment for installing the sensors is hundreds of millions of yen on the whole, but the company expects to be able to have a return on this investment in three years.

Panasonic's power saving measures for this summer are mainly for the short term, but the company intends to keep tackling the issue by continuing to make improvements to the measures after the next fiscal year.

Bibliography

1. Nihon Keizai Shimbun, *A 15% conservation in power usage without reducing production output,* May 14, 2011.
2. Panasonic, *Panasonic's engagement with the power saving initiative.* http://panasonic.co.jp/eco/setsuden/ (20120930)

PART III

Case Studies of Super Effects in Marketing Domains

Section 1: Size Changes

32

The Process of Organizing Information

Atsushi Tsujimoto

The Target is "Ms. A, a 25-Year-Old Single Female Office Worker Living in the City"

The brand *Francfranc* (*BALS CORPORATION*, Shibuya-ku, Tokyo, President Fumio Takashima) targets consumers who match a particular consumer image championed by the company. Among marketing achievements made with such a premise, this company's achievement is superlative. *BALS CORPORATION*, which manages this brand, is classified as a major company dealing in general merchandise, furnishings, and the production and sales of items related to interior decor. The company boasts a consolidated sales of 33 billion yen (as of the January 2011 period) for the domestic and foreign operations of its group. *Francfranc*'s image of its target consumer is "Ms. A, a 25-year-old single female office worker living in the city". In particular, the company positions itself as the answer to such a consumer's intense longing for living in a residence with stylish interiors like those that appear in magazines.[1]

Twenty percent of their merchandise mix is comprised of furnishings and fabric (textile and woven fabric products) while general merchandise accounts for 80%. The company is said to have mainly young women as its followers, who follow enthusiastically. At present, their marketing has evolved to incorporate more sensitivity: Instead of being focused on age, the marketing places an emphasis on appealing to the young at heart and others who are sensitive to the trendy styles.

Francfranc's sales accounts for approximately 86%* of the company's consolidated sales (as of the January 2011 period). I shall elaborate below on the business development peculiar to this brand.

The Motto of Product Development and Marketing Style

> "We update the lifestyle and mindset of consumers with *Francfranc*'s product concept. We do not propose to make up for some functions that are lacking in their lives. While we used to listen to what our customers had to say a long time ago, ever since we began to think that we should be the ones leading, we stopped listening (to them now)".
>
> (*BALS CORPORATION*, President Fumio Takashima)

Mr. Takashima tells his employees to "play often". By that he means "know," "watch," and "eat". "To make your everyday life substantial, you need to dig deep and go beyond the ordinary. We as a company should have a deeper understanding of film, art, sports and cuisines than customers. It's our duty, I think," he said, recollecting the time he elaborated the plan for the brand called *WTW*.[3]

"Although I had been wishing to do a natural brand for several years, I never quite got around to it. However in 2009, when I was in Kamogawa, Chiba, waiting on a surfboard for a wave, I was struck with this sensation and began to wonder, 'What's going on here? What's this feeling I'm having? I'm feeling good even though I'm not riding on a wave right now. I'm feeling this power welling up from underwater'. Hey, that's it, I thought. This is the feeling! All I have to do is express a style that reflects this calmness and the idea of surfing in a natural way. That'll do it," Mr. Takashima thought on top of a wave.

The everydayness that was sensuously absorbed in a state of play emerged as an image that welled up in his mind, which in turn gave rise to a brand.

Francfranc's proceeds were on a rising trend until 2009 and began to decline from 2010. "The whole company became overconfident and slack. It became proud. We were beginning to have less original ideas," Mr. Takashima recalled. "In that year, I kept saying, 'Don't make any products. Don't open shops. If you've got time on your hands, go get some sleep. Get out and have fun somewhere'". These were instructions aimed at honing his employee's sensibilities toward everyday life. In other words, he was issuing a mandate for them to go out and search for a new personal worldview or a concrete sense about some aspect of living that can translate into seeds for product development. It can be seen that Mr. Takashima was harboring a sense of crisis, fearing that the worldview of his employees was becoming vacuous.

The brand perceives the latent needs of consumers and concentrates on helping to realize them. It attempts to make consumers think, "How did you know what I want?", "*Francfranc* is truly marvelous". By sustaining such an effort over time, Mr. Takashima says a company will become capable of leading the market.

However, strictly speaking the company's target is "Ms. A, the 25-year-old female office worker" and the question for the company is whether it can swiftly reflect Ms. A's genuine motivations hidden deeply behind the signals she transmits. Mr. Takashima emphasizes that this is precisely the brand's pulse, or *raison d'etre*.[5] In effect, the company strives to create products that are strictly suited to Ms. A's sensibility by having each and every employee hone his or her awareness of everyday life and accumulate the images perceived from their sharpened awareness. In effect, the company repeatedly engages in the task of reconciling their products with images.

"The power of cashmere to warm your heart and also your romance". This was the copy Mr. Takashima created for a cashmere blanket his company offered. It appeared on explanatory labels. For a potholder, the copy read, "I hold your hand", and for a bar of soap, the copy read, "Lather, lather and double your girl power".

"I did not wish to just sell things," Mr. Takashima said. "I wanted to convey the scenes that featured the products in them and scenes that the products themselves bring about as well. Instead of only writing down descriptions like 100% cashmere or texts that explain functions, I wanted to also make it possible to convey scenes that gave consumers an idea of what they could do with the product if they made use of it".

It can be understood that the company prioritizes the creation of images that allude to how particular moments or scenes in everyday living could become more colorful just by using a particular product, which can add a sense of opulence to the image of a customer's daily life. The objective of eliminating the inconveniences of daily living is secondary to helping customers fill up the empty spaces in their hearts.

Starting Point: Understand the Zest of Everyday Living from a Physical and Emotional Point of View

WTW, a brand whose concept is "natural," was formed when a certain image emerged in Mr. Takashima's mind as a palpably physical and emotional response to being in "a play-space". The novelist Ryu Murakami once likened him to "a person who thinks with his body". These words are indeed appropriate and are applicable in describing the staff members running *Francfranc* as well.

In a sense *Francfranc* is realized by perceiving the essence of everydayness through the lens of emergent physical sensations and emotions and reflecting this essence in product development.

Ideally, the process would entail the product developer commercializing an exuberant mental image that stems from moving emotional experiences of everyday life and making consumers who pick up a product perceive the same image (to the extent possible). The issue then becomes how similar the image reproduced in the mind of the consumer will be to the image harbored by the developer, or in other words, how similar the image will be as it is adapted and integrated into this person's life image. Below, you will find a summary of *Francfranc*'s approach toward product development and sales.

1. Commercialize the mental image/worldview that arises as a physical and emotional response to experiences that transpire in everyday living.
2. At stores, prepare materials that illustrate a worldview associated with a product in such a way that the mental imagery that arises in the mind of a customer who sees this product or holds it is similar (as much as possible) to the imagery the developer of the product harbors for the product.

Products created under the brand called *Francfranc* are certainly functional but they are more valuable for many other reasons.

I believe the greatest value added feature of the *Francfranc* brand can be found when it evokes something special in the customer's heart; some distinctive worldview/emotion associated with its products when they become integrated into his or her life.

Notes

*Based on the data found in the section titled *Sales by brand* in the report *Highlights of financial affairs and results* (as confirmed on June 25, 2011) http://www.bals.co.jp/ir/achivement/index.html

**The source for the comments made by Mr. Fumio Takashima and Mr. Ryu Murakami is source No. 6 below.

Bibliography

1. Fumio Takashima, *Thoughts I had as I managed Francfranc,* Keizaikai, 2008, p. 30.
2. Fumio Takashima, *I don't want employees who don't play — Job performance is determined by the amount of time you find enjoyable,* DIAMOND, Inc., 2010, pp. 87–88.
3. TV Tokyo Corporation, *Woman Mania! The way Francfranc grasps consumer needs,* broadcast on "Cumbria Palace" on March 10, 2011, Thursday, from 10 pm–11 pm.
4. BALS website. Available at http://www.bals.co.jp/ (20110310)

33

The Learning Process in Product Development

Atsushi Tsujimoto

P&G's Superb Sales Performance

I will elaborate on the subject of team learning that takes place in the product development process at *P&G Japan* (Procter & Gamble Japan Co., Ltd., Higashinada-ku, Kobe-shi, President Kazunori Kiriyama) that facilitates superior business performance. Specifically, I will illustrate this process from the perspective of the Hierarchal Autonomous Communication System model (HACS).

Established in 1837 in the U.S.A., *P&G* is the world's largest public consumer goods manufacturer and it began its operations in Japan in 1973. The company handles cosmetics, detergents, beauty consumer electronics, healthcare products (including digestive medicine and water purification systems) and products related to pet care, baby care and family care.

Its net sales amount to approximately 78,900 million dollars, which is comparable to the net sales of *Toyota Motor Corporation* (Toyotashi, Aichi, President Akio Toyoda). As a business enterprise that makes and sells general consumer goods, it is the largest in the world, continuing to demonstrate superlative sales performance.

The Learning Process in Product Development — Home Visit Interviews

Renowned as a company that carried out marketing research ahead of others in the American industrial world, *P&G* thoroughly investigates what kinds of products consumers like. Just as the organization's slogan, "The consumer is boss"[1] alludes, the company believes that the foundation of all manufacturing lies in understanding that "everything is for the consumer", that "the consumer is the starting point", and in thinking about "making the consumer happy".

This corporate stance revolving around the consumer is also indicated by the large amount of marketing expenditure the company incurs (approximately 35 billion yen a year (for about 5 million people in about 100 countries).[2] To investigate consumer needs, the company attaches great importance to "fact finding visits" which involve employees of the company actually visiting the homes of consumers. In particular, the R&D staff members who are directly involved with product development visit sites (homes of consumers) and probe the consumers to gain a sound understanding of their spending habits from the perspective of depth psychology. In effect, the company is practicing a distinctive approach (a learning process) in which the "maker" investigates to understand consumer tastes. In this approach, observing how the consumer-monitor generally spends his or her days and having casual conversations with him or her is considered vital.

The washing detergent "*Ariel Revo*" whose catchphrase is "It will make it hard for food stains to stick" was born from such a fact-finding visit. A mother was said to have been "frustrated" by her failure to get rid of food stains stuck on her child's clothes. "*Ariel Revo*" was a product that was created in response to the question of whether such a mother's stress could be relieved in some way. The R&D head of the company describes the basic line of thinking behind the kind of product development that draws on consumer depth psychology as follows.

"You really have to visit the scene of action to gain a firm understanding of things...and in this case I believe the scene of action is "the home". Although we don't hear what people are dissatisfied about that much by just asking them and although interviews can elicit responses such as 'I feel slightly frustrated' or 'slightly troubled', we do gain considerable insights from just observing them anyway so we make it a point to visit them at their homes to let us make observations".*

The *Ariel Revo* contains an ion polymer that was developed by the company. By washing with this detergent, clothes become more resistant to food stains since their surfaces become coated with ions. For this reason, *Ariel Revo* became a revolutionary product that championed a new approach to laundry known as "preventive washing".

A HACS Model Interpretation

To understand specifically what the above-mentioned process of carrying out fact-finding process (the learning process) is about, it is significant to explain it within the framework of information science, or particularly within the framework of Fundamental Informatics, using its analytical framework known as HACS. Doing so will enable us to understand the company's approach on a micro level.[3]

HACS is one form of a verification model and it is a methodology that supports Fundamental Informatics — a field of cutting edge research that considers information from a multi-faceted perspective. In brief, this is a methodology that decodes an endemic information phenomenon forming the premise of the operation of a system by observing and describing the phases of communication.

While *P&G*'s fact-finding process (learning process) is an operational phenomenon of one particular system, there is a need to first of all basically explain what kinds of information phenomena (communication phenomena) are being allowed to emerge and develop. Therefore, I will elaborate on this process within the framework of the HACS model.

The HACS model mentioned here can be considered to be a complex model in which two subjects are combined. The first subject is the consumer monitor who comes under society's influence, contemplates it and then acts. The second subject is the product development manager who observes this monitor and attempts to draw out valuable information (through fact-finding). Henceforth, the consumer monitor will be referred to as A and the product development manager as B.

Observing B, A ascertains there are situations when B feels "slightly frustrated" or "slightly troubled." It is believed that when B carries out routines of everyday living (i.e., actually carrying out washing) at the time of the on-site research, B's feelings of becoming slightly frustrated or slightly troubled become amplified. These considerations are verbalizations of A's perceptual world and are to be understood as "subjective utterances/first person descriptions". For this reason, third parties tend to find them difficult to understand.

B observes A's utterances (or A's conduct) and takes notes (acts for description). Since B is observing the scene in which A is actually washing at home, B can understand A's perceptual world (the world of A's sense of inconvenience) at a concrete level. In this way, B attempts to understand A's perceptual world (as much as possible).

"These actions form what is referred to as a (para) objective world/third-person description" and are also actions that form "objective information" that can be shared among the persons involved (as much as possible). That concludes my analysis based on the content of citation No. 2. What follows are my suggestions based on the HACS model for communication conducts that should be carried out in the course of developing a product.

It is necessary for B to place the above-mentioned process (the process of converting information from the first person description to the third person description) on the level of the abstraction process, which can be described as the process of arriving at a "meta-third person description". To fit the descriptions into the framework of the marketing logic of the product development team to which B belongs, it is necessary to convert the nature of the descriptions. (Incidentally, this marketing logic entails criteria based on scientific premises with a high level of abstraction (or a high level of relatability) that include assumed viability of new products, their profitability, their market potential and their impact). At this point, "the (para) objective world" approaches extremely close to "the objective world". In conclusion, acts of description can be understood to unfold in the order of numbers one through three as described below.

1. "First person description" (make the consumer monitor to first of all let the subjective world emerge)
2. "Third person description" (Formation of the (para) objective world (attempt to make descriptions recognizable to third parties as much as possible)
3. "Meta third person description" (Formation of the meta (para) objective world) (aligning descriptions to the marketing logic in play)

I believe it is necessary to note the dynamism that sees the layering of the substitutive actions of numbers one through three that change the value of recognizable information through "observations/descriptions".

The product development manager who conducts hearings (the observer) is unable to decode the consumer monitor's "subjective world/ perceptual worldview) in its entirety nor does he or she need to. I believe it will suffice if even just a fragment of the consumer monitor's subjective world/perceptual worldview becomes a resonant piece of information in the framework of the relevant product development logic so that it ties into the development of a new product.

Note

*Comments made by the company's employees on the on-site research visits and their research processes are all drawn from the contents of citation No. 2.

Bibliography

1. Makoto Takada, *P&G's communication technique is the company's consistent strength — Communication supports the company's 170-year growth,* Asahi Shimbun Publications, 2011, pp. 66–68.
2. TV Tokyo, *What is a true global company!?,* Cumbria Palace, broadcasted on June 9, 2011 (Thurs), from 10pm to 11pm.
3. Toru Nishigaki, *Fundamental Informatics — For the Living Organization,* NTT Publishing, 2008.
4. P&G Japan website. Available at http://jp.pg.com/index.htm (20110609)

34

Signs of Explosive Expansion in Global Geothermal Power Generation

Toru Fujii

The Scale of Global Geothermal Power Generation and its Existing Circumstances

In Japan, in the wake of the nuclear accident of the Fukushima Nuclear Power Plant, the establishment of an energy policy that will serve as an alternative to the nation's nuclear energy policy has become an urgent challenge. Meanwhile, the use of geothermal energy for power generation is receiving renewed attention as the price of fossil fuel soars suddenly and countries all over the world are investigating ways to reduce emissions of greenhouse gases and their dependence on oil.

According to the Earth Policy Institute (Washington, DC), there are vast reserves of thermal energy existing in the earth's crust, which is approximately 10km thick; it is known to amount to 50,000 times the amount of energy that could be derived from combining all the existing resources of oil and natural gas. The number of active volcanoes in Japan is 119, making it the third among countries with the most number of volcanoes after the United States and Indonesia. In addition, in proportion to this, geothermal energy resources amount to 23,470,000kW (kilowatts) of energy in the top three countries, overwhelming its use in other nations.

A revision of Stefansson (2005), showing positive correlations between the number of active volcanoes and the quantity of geothermal energy resources of the world's major geothermal energy resource-rich nations. As a result of entering new evaluation results in the quantity of Japanese geothermal energy resources, you can now see that the three major countries that overwhelm others in terms of geothermal energy resources are Indonesia, the United States, and Japan.

These are outputs obtainable from depths considered to be reasonable given the current technology and economic efficiency, and if strides are made in power generation technology along with cost reductions so as to make it possible to drill down to deeper depths, the institute believes the attainable output will become much larger.

Furthermore, a geothermal power generation facility currently yields 540,000 kW, serving 0.2% of annual demand. Consequently, if the facility becomes 150 times larger than it is now (to produce 81 million kW) it is thought that it will then reach the scale of a nuclear power generation facility (30% share). Although serving Japan's total electricity demand with only geothermal power would require a geothermal power generation facility 500 times larger to produce approximately 270,000,000 kW (270,000 MW), expectations for geothermal energy remain high since this form of energy's reserve never has shortages.

Significance of Geothermal Power Generation

Power generation *via* the use of geothermal energy as a source began in 1904 in Larderello, Italy and today this form of power generation is currently being carried out in 24 countries. Among them, five countries get more than 15% of all their domestic electric power through geothermal power generation. By early 2008, the total capacity of the world geothermal power generation facilities combined had exceeded 10,000 MW (megawatts), producing an electrical output that meets the demand of 60 million people, which is comparable to the population of the United Kingdom.

The reserve of thermal energy existing within the earth's crust, which is approximately 10km thick, is immense and is known to be 50,000 times the amount of energy that could be derived from all the resources of oil and natural gas combined. This energy is derived from the core of the earth and the decay of natural radioactive isotopes such as uranium, thorium and potassium.

In particular, countries located in the Circum-Pacific Volcanic Zone (an area surrounding the Pacific Ocean consisting of a high level of volcanic activity) are rich in geothermal energy. These countries include Chile, Peru, Mexico, the United States, Canada, Russia, China, Japan, the Philippines and Indonesia.

The number of countries with geothermal resources that completely fulfill their respective domestic electricity demand is 39, whose populations in total exceed 750 million.

To achieve geothermal power generation, an underground reservoir (cistern) containing deposits of thermal water and steam was usually required to operate condenser turbines. However, power generation at temperatures far lower than conventional levels has been made possible, thanks to a new technology that uses a liquid with a low boiling point through the application of a closed heat exchanging system. This epoch-making advancement has allowed even Germany, a country hitherto unassociated with geothermal resources, to generate geothermal power.

The Future of Geothermal Power Generation and its Use

The strengths of a geothermal power station lie in the fact that it leaves a low carbon footprint, its fuel cost is low, and that it can generate electricity with the use of local energy resources. Moreover, the base load electric supply that is required on a constant basis can be supplied 24 hours every day. In addition, there is also the merit of having no need to keep a supply of electricity on reserve or set aside one for use as an emergency power source.

The United States generates the most amount of geothermal power in the world. The total volume of geothermal power generated as of August 2008 in the seven states: Alaska, California, Hawaii, Idaho, Nevada, New Mexico, and Utah amounted to approximately 2,960 MW.

In California, the facilities require a capacity of 2,555 MW, but approximately 5% of electric power generated derives from geothermal energy, making this percentage larger than in any other country. Most of these facilities are found in a complex of geothermal power plants called *The Geysers*, which is located in northern San Francisco where geologic activity is vigorous. With the passing of the Energy Policy Act of 2005, geothermal power generation became tax deductible as an incentive under the policy for carrying out renewable energy production. For this reason, in many markets in the western part of the United States, the cost for generating electricity with geothermal energy as a source has become equivalent today to the cost for generating electric power through the use of fossil fuel as a source.

For the future, as an alternative to fossil fuel, which has a high carbon footprint and whose price fluctuates wildly, renewable energy sources that are highly cost effective with a low-carbon footprint are attracting the attention of nations. Consequently, amid such a situation, geothermal power generation is poised to become mainstream at once, even though its consumption is only a little at present. In addition, as part of a plan to reduce global carbon emissions by 80% by 2020, the Earth Policy Institute has declared that it will aim to help generate 200,000 MW of geothermal power.

Bibliography

1. Earth Policy Research. Available at http://www.earthpolicy.org (20110529)
2. Tadashi Ono, *Regarding the geothermal resource/Regarding geothermal energy*. Availble at http://geothermal.jp/power/modules/pico/index.php?content_id=2 (20110529)
3. Tokyo Electric, Public Relations Department of the PR Planning Group, Hirofumi Muraoka, *Establishment of the Energy Policy Act of 2005 in the United States,* TEPCO REPORT, Tokyo Electric, Vol. 112, October 2005, pp. 12–13. Available at http://www.tepco.co.jp/company/corp-com/annai/shiryou/report/bknumber/0510/pdf/ts051004-j.pdf (20110529)
4. Hirofumi Muraoka, *Promoting Geothermal Energy Development as a Paradigm Shift,* Gate Day Japan Symposium, Part III of AIST's lecture materials on their research, National Institute of Advanced Industrial Science and Technology, August 5, 2009, p. 16. Available at http://staff.aist.go.jp/toshi-tosha/geothermal/gate_day/presentation/AIST3-Muraoka.pdf (20110529)

35

Smart City/Smart Community Markets

Toru Fujii

The Trend in Smart City/Smart Community Markets

Urban development around the world is undergoing great transformations. Firstly, there is a shift toward creating "low-carbon cities" by reducing CO_2 emissions to deal with global warming. Secondly, China and other emerging nations in Asia and South America are seeing higher population concentrations in their cities and the creation of more metropolises to meet the needs of their economic development. Thirdly, cities in advanced countries, including Japan, are aiming to realize health-conscious urban developments that also aim to close the generation gap by developing urban areas revolving around medical and welfare facilities to deal with issues related to the graying of society. Fourthly, there are cities seeking to develop safe and reliable urban constructions to protect towns and regions from the impact of natural disasters, such as earthquakes and tsunamis.

In recent years, emerging nations including those in Asia have seen rises in population inflows to urban areas and increases in traffic densities due to sudden industrial expansions of their economies. This situation has been problematic, intensifying breakdowns in urban functions and making industrial structures more complex. Since such cities are unable to cope with such problems under conventional urban structures, it is thought that demand will rise increasingly for carrying out urban expansions, redevelopments, and radical reinforcements of urban functions.

The total amount of investments around the world for urban developments in 2005 was approximately 230 trillion yen. By 2020, this amount is predicted to escalate to approximately 360 trillion yen. Expansion of demand for these investments has been substantial in the newly industrializing regions of Asia, such as China, India, and Vietnam, where urbanization schemes are rapidly advancing. In India, where there were 42 metropolises of more than one million people in 2008, it is predicted that 68 metropolises of such a size will be built by 2030.

In addition, it is assumed that there are six megacities of more than 10 million people and it is anticipated that demand for urban developments in mainly emerging nations will see a rapid escalation. Furthermore, expectations are high for business developments in various service sectors, such as livelihood support services and maintenance services, including maintenance management and large-scale renovations for urban functions in next-generation eco-conscious "smart cities/smart communities".

In the world, 300 to 400 smart city projects are underway. As shown in Fig. 35.1, if we combine all the markets associated with the energy sector — comprised of the market for constructing power collection networks, the market for frequent uses of renewable energy such as photovoltaic power and wind power, the market for mass installations of

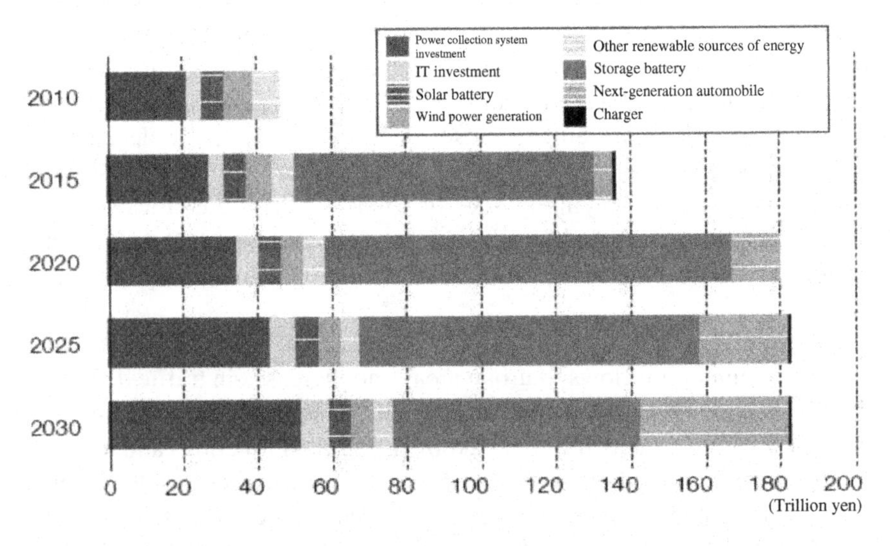

Fig. 35.1 The global market for smart cities/smart communities

Source: Yosuke Mochizuki, "The smart-city market size will reach 180 trillion yen by 2020"

storage battery systems, and the market for introducing next-generation automobiles — we will have a total market size of 180 trillion yen by 2020.

The market for smart cities and smart communities based on these energy sources was 45 trillion yen in 2010, but this figure is expected to reach 135 trillion yen by 2015, and by 2020 it is expected to quadruple from the current amount to 180 trillion yen. In effect, the market is predicted to achieve dramatic growth over the coming 10 years.

With smart cities, we can expect expansions in areas other than the energy sector, such as in the businesses of water infrastructure constructions and waste disposal treatment, the business of constructing smart houses and smart buildings, sales of smart consumer electronics, and in the businesses of the energy service industry. However, the scale of investment in just the energy sector amounts to the same level as required for all these sectors. If we examine by region, we see that North America and Europe were leading the world at the outset, but after 2015, China will come to lead. Beyond 2020, India will see growth.

While the smart city market is expected to see substantial growth in this way, if we examine only the market for investments in energy, we see that its growth will remain at the same level after 2020. This indicates that the ten-year period from 2010 through 2020 is a period of opportunity for launching startups, and that the focus of growth thereafter will shift toward new services that offer applications. This leveling off is mainly attributable to anticipated photovoltaic power generation, which is a renewable energy, and to an anticipated sudden drop in the unit price of storage batteries installed for stabilizing systems.

Japan's Governmental Measures

In the cabinet decision entitled, "Realizing New Growth Strategies 2011", the cabinet concretely shows in the "21 National Strategy Projects" the main results and issues it anticipates for 2011.

Namely, the cabinet proposes the introduction of the smart grid, the establishment of rules for power system operations, the establishment of measures for extending system interconnections, upgrading electric power systems, carrying out proof of concept enterprises for a next-generation energy society. In addition, the cabinet also proposes to offer systematic support for an electric power industry subcommittee that will take into account the results of study meetings for considering next-generation power transmission and distribution systems and smart meter systems. Moreover, the cabinet is also attempting to promote regulatory reforms and zonings to establish locations for wind power generation, geothermal power generation, and wind power generation at sea (with the cooperation of the fisheries cooperative association).

The Strategy of *Panasonic*, a General Home-Appliance Maker

This Japanese general electrical equipment manufacturer is deploying a strategy that has at its core the Home Energy Management System (HEMS) and the Building Energy Management System (BEMS).

Specifically, the strategy calls for developing a business that accommodates the entirety of a product's life cycle, which encompasses the following phases:

1. The startup phase in which plans and designs for urban-development related fields are developed, designed, and executed;
2. The product system delivery phase; and
3. The operations management phase.

I will now introduce the measures taken by the general home-appliance maker, *Panasonic Co., Ltd.* (Kadoma-shi, Osaka, President Fumio Otsubo). In its 2010 management policy, *Panasonic* describes its smart city business as an offer of a "one-stop energy solution" service, specializing in providing one-stop energy solutions for houses and buildings that other companies are unable to provide.

As a general home-appliance maker, *Panasonic* is attempting to spread its HEMS-based solutions from individual homes and buildings to entire blocks and local communities to pave the way toward building community grids. Consequently, by reinforcing the value of these solutions by associating them with "energy creation," "energy storage" and "energy conservation," and by reinforcing their global salability, the company attempts to establish its line of energy systems as its flagship business.

The characteristic feature of *Panasonic*'s strategy is to initially develop businesses that have customers purchasing the company's consumer electronics and then to carry out from outside the customers' domain the services related to "energy creation," "energy storage" and "energy conservation" for those products owned by the customers. For the future, the company is considering information services that chiefly aims to offer "solutions" achievable by these consumer electronics while aiming to expand services such as leasing and renting. In effect, the company is considering business developments that go beyond its conventional framework.

Bibliography

1. Fumio Otsubo, *Panasonic 2011 management policy — Management vision and strategy*, Panasonic, April 28, 2011. Available at http://panasonic.co.jp/ir/vision/pdf (20110526)
2. Cabinet decision, *Realizing New Growth Strategies 2011*, The National Strategy Chamber of the Cabinet Secretariat, January 25, 2011. Available at http://www.npu.go.jp/policy/policy04/pdf/20110125/20110125_01.pdf (20110526)
3. Cabinet decision, *Summary of the 21 National Strategy Projects — New growth strategies? Scenarios for revitalizing Japan?* The National Strategy Chamber of the Cabinet Secretariat, January 25, 2011. Available at http://www.npu.go.jp/policy/policy04/pdf/21project.pdf (20110526)
4. Panasonic website. Available at http://panasonic.co.jp/ (20110526)
5. Yosuke Mochizuki, *The smart city market will be worth 180 trillion yen by 2020*, Smart City Frontline, Nikkei BP, November 26, 2010. Available at http://eco.nikkeibp.co.jp/article/report/20101122/105335/ (20110526)

36

The E-Learning Market's Growth Potential

Akira Ishikawa

The Utilization Rate is Only 4%

In this section, I will elaborate on e-learning as a new business. It is among those businesses that are resilient to the impact of recessions and its main fields are education and training.

When we survey the history of e-learning, we can see that it has its origins in the field of Computer Aided Instruction (CAI), which is mainly oriented toward serving the manufacturing industry. This CAI-based form of learning was already in use since the end of the 60s in the departments of science and engineering of companies and universities, and thereafter went on to form the foundations of Computer Based Training programs (CBT), Web Based Training programs (WBT), and Web Based Conferences (WBC).

Since then, CAI came to be known as e-learning as it came to encompass asynchronous or on-demand WBT forms of learning, synchronous or real-time forms of learning using video conference systems through the use of satellite communications or the Internet, self study forms of learning using CD-ROMs, and the mobile learning platform using handheld units.

According to the white paper on e-learning by the Ministry of Economy, Trade and Industry's Commercial Affairs Information Policy Department's Section of Information Processing Promotion, e-learning became widespread around 2000. However, according to a 2007 survey, the rate of usage for e-learning programs was merely 4.0%, as shown in Fig. 36.1

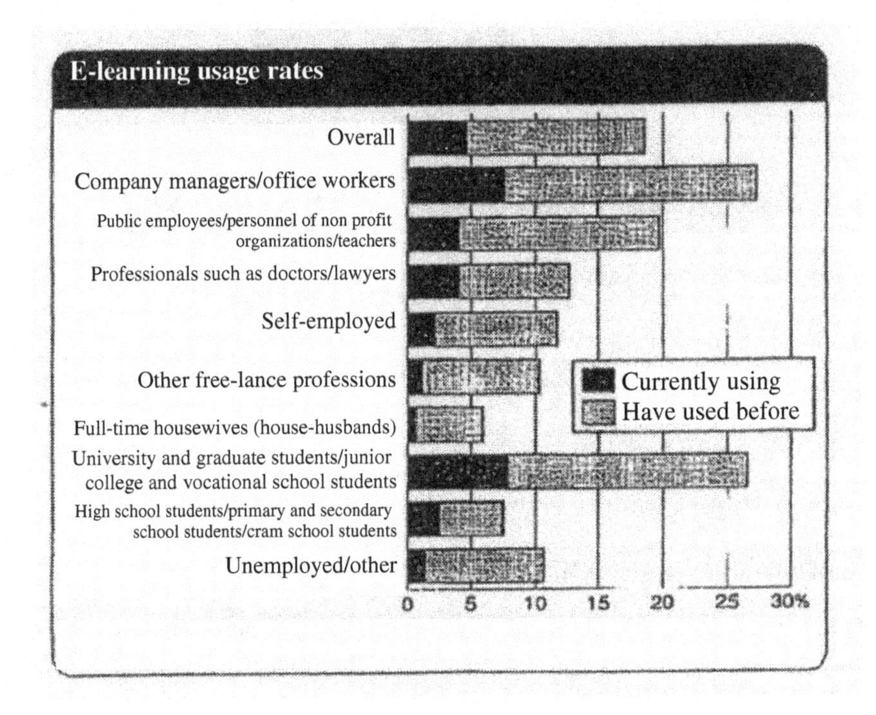

Fig. 36.1 E-learning usage rates

(*The Seventh Market Survey on Broadband Content Usage*, Joint research between Mitsubishi Research Institute [Chiyoda-ku, Tokyo, President Kyota Omori] and *NTT Resonant Inc.* [Minato-ku, Tokyo, President Takao Nakajima]). Although this figure leaps to 18.6% if we include people who have used e-learning in the past, nearly 15% of the respondents appear not to be taking advantage of e-learning programs currently.

The population of broadband users is estimated to be approximately 50,000, so around two million people are considered to be e-learning users.

Two years prior to the above-mentioned survey, in the 2005 edition of the information and communication white paper (Ministry of Internal Affairs and Communications), Japan's e-learning utilization rate (according to a January 2005 survey) was reported to be much lower at 3.1%. The survey also reported that the corresponding rate in the U.S. was far more at 24.0% and in Korea it was even more at 25.0%.

Fifth Place in Asia

Even in terms of e-learning preparedness among Asian nations in the training, industrial, and governmental sectors, Korea ranked at the top and was followed by Singapore in second place, Taiwan in third place, Hong Kong in fourth place, and finally Japan in fifth place. One of the major reasons why the usage rate in Japan was so low compared to the United States and Korea is because compared to Japan, governments of both countries put far more effort into spreading e-learning.

In the case of the United States, to eliminate dropouts, the "No Child Left Behind Act" was passed in January 2002. This act has the substantial goal of making all students attain a prescribed standard of education by 2014. To guide the achievement of this aim, the act clearly states that the government will offer online educational opportunities for all students, making them a compelling driving force behind the spread of e-learning.[1]

Korea also Backs Online Education

Meanwhile in Korea, with the "E-Campus Vision 2007", the government applied e-learning to regular high-school curriculums and built user-centric learning systems. But that was not all. Through e-learning, the government also attempted to revitalize credit exchanges with other universities and to offer consummate high quality on line educational programs for local communities and regional industries.

In addition, the government has officially recognized the Cyber University, which offers only online courses (Higashi-ku, Fukuoka-shi, President Moichi Hirasawa). This university has reportedly grown to a sizable level today and is enjoying an enrollment of over 67,000 students.

Even in China's case, there were similar activities, and in January 2005, the Education Department announced the "Trial Operation for the Partial Integration of Online Curriculums". For the first trial held in May of the same year, there were only 300 participants, but for the official online trial held in 2006, it has been reported that the number of participants amounted to 280,000.

However, this is not to say that Japan has not experienced desirable developments of its own. According to the survey conducted by *Yano Research Institute Ltd.* (Nakano-ku, Tokyo, President Takashi Mizukoshi), the total scale of e-learning programs, ranging from educational and training services for corporations to those using educational software that run on gaming devices like the "*Nintendo DS*," has apparently reached 134,100 million yen.

And whereas the market size increased by 7.8% from the previous fiscal year, the market for e-learning services for individuals that make use of PCs and the Internet increased by a whopping 25.7% while game console-based e-learning programs grew by an extremely surprising 41.6%. For these reasons, we can see that a grassroots level e-learning movement is also beginning to catch on in Japan.[2]

Seeing that the largest company in this sector has started to take great interest in this field, and with expectations rising for the government to take more active measures, my hope is that the educational tool, that is e-learning, will serve as a stepping stone toward the improvement of Japan's standard of education, which has been declining.

Bibliography

1. NTT Data DIGITAL GOVERNMENT Henshukyoku, *E-learning policy of the U.S. Department of Education* (20110530)
2. Terunobu Kinoshita, "*The Age of the Digital Natives,* Toyo Keizai Shinposha," 2009, p. 108.

37

Expansion of Management Consulting Project Scenarios *via* Information Dispatches

Hiromichi Yasuoka

The Necessity of Transmitting Information in the Management Consulting Business

Management consulting in brief is about precisely grasping problems faced by companies and providing solutions to solve their problems after having taken into account the context of various kinds of business sectors. There exist various management consulting fields, ranging from consulting on strategies to consulting on business operations and systems.

Since these days even salespeople within companies need the ability to provide consulting services, many introduce themselves as a "so and so consultant". Twenty years ago, there were only a limited number of companies that recognized this value, and even in those firms, there were a limited number of personnel who were called "consultants". Consequently, these days, it has become necessary for consultants to differentiate themselves. Of course, there are academic degrees that serve as a gateway for professionals to become consultants, such as the Master of Business Administration (MBA) degree and other degrees that serve as qualifications for providing consulting services for small-and medium-sized companies or for providing financial analysis. However, these degrees alone do not easily lead to consulting contracts.

Consequently, tailoring and transmitting information for the media becomes necessary. In particular, when consultancies convey their field of specialty to the media more (such as to various newspapers and magazines), such information dispatches often get picked up as various news stories, unlike TV commercials simply aimed at the general public. By doing so, consultants inquire from various businesses increase as well as from the media. Consequently, consultants will have more information, and the level of their expertise will rise. When such a virtuous cycle occurs, requests for hiring these consultants see an increase.

Nomura Research Institute (Chiyoda-ku, Tokyo, President Tadashi Shimamoto) is a consulting firm and a think tank (while being a systems company as well). The company particularly attaches great importance to information dispatches. These days, even foreign consultancies such as *A.T. Kearney* announce and appeal various provisional calculations to enhance the perception of their expertise in their field. By doing so, consulting firms can amplify their presence in their areas of expertise.

In this day and age of highly advanced digitization of information, using ads to advertise a company's expertise may not prove to be effective as desired in the end. It is therefore important to have a third party also to advertise. To that end, it has become necessary to carry out mediative profiteering — which is the tactic of introducing a problem and to be rewarded for solving it — through information dispatches. This idea is illustrated in Fig. 37.1.

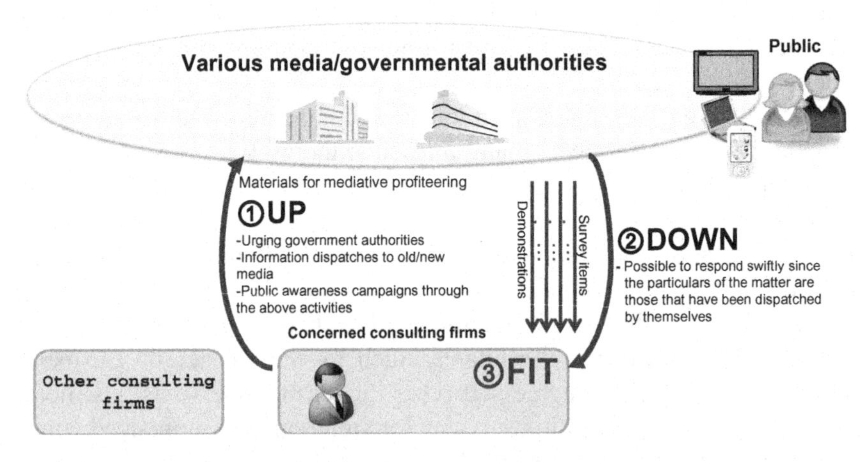

Fig. 37.1 Information dispatches for mediative profiteering

Source: Author.

In addition, there have been cases outside of the consulting sector that have seen the development of businesses from an upstream process in the way mediative profiteering works. A case in point is the eco-points scheme implemented by the Department of Environment, Ministry of Economy, Trade and Industry, and the Ministry of Internal Affairs and Communications. The advertising agency *Dentsu* was entrusted with this scheme and was able to connect it with its main business, which is advertising.

The Effectiveness of Mediative Profiteering Through Information Dispatches for the Media

I am currently working at *Nomura Research Institute* (NRI). I had been with NRI originally, but I became involved with the launch of an Internet business for an entertainment firm during the period of the Internet bubble and thereafter worked at a foreign (accounting-related) consulting firm. However, I ended up returning to NRI. Consequently, since 2003, I have been striving to receive consulting project contracts, carrying out information dispatches from around 2005. Since then, I have succeeded in promoting myself as an expert in my field, seeing a large increase in the volume of my contracts. Of course, however, since the consulting business serves as an indicator of a business cycle's future course and since it is also a labor-intensive industry, orders for its services come in waves. However, my efforts produced a clear difference.

Now let us look into a specialized field. For example, in the field of Customer Relationship Management (CRM), *Accenture* was one step ahead of its pack at the outset, publishing "CRM — Customers Are Just There" and being invited to attend various consulting-scenario competitions.[1] However, as a CRM measure, it is always necessary to examine point programs. Pertaining to this field, *NRI* published a report titled, "Enterprise Currencies of 2010 — The Point Economy of the Googlezon Age" and attracted attention for conveying the idea that points issued by companies can be used as pseudo currencies, (enterprise currencies).[2] In these enterprise currencies, electronic monies were also included and since these fields were also seeing the implementation of system development, *NRI* went ahead one step further. The company then went on to enlarge this domain even more, taking it up also as a business that made use of IDs and as a business related to payments, making *NRI* considerably advanced in this field.

In the field of consulting, it is necessary to clearly convey your authority and leadership by going one step (not two steps) ahead of the customer and talk about the essentials from a big-picture perspective. To this end, since mass-market books are easy to understand, it is easy to have readers recognize your authority and such a perspective.

As mentioned above, *NRI* was able to lead in certain areas. However, since we are in an information society, there is a likelihood naturally that *NRI* might see its situation reversed immediately unless it continues to carry out information dispatches and work on improving its level of expertise. Nevertheless, *NRI*'s case is exemplary.

Due to information dispatches designed for mediative profiteering, the contribution I made to the consulting firm in terms of receiving orders has exceeded 100 million yen many times, as shown in Fig. 37.2. Incidentally, among the managerial class, which comprises one fourth of NRI's workforce, only 10% of them are consultants whose contributions have exceeded the 100 million yen mark. Moreover, those who have made contributions of that level over four times are extremely few. In other words, this is the same thing as saying the figure is one tenth to the fourth power and a simple calculation shows that the probability of achieving such a result is 1/10,000. Of course, the number of highly achieving consultants are limited to a select few so the figure doesn't amount to 1/10,000, b ut there is no question that I have achieved a super effect compared to the time I began and compared to my competitors.

Fiscal year	Order contribution amounts (estimates, including outside of the consulting sector)	Notes (major information dispatches, etc.)
2003	20 million yen	Business starts in June
2004	30 million yen	-
2005	100 million yen	Information dispatches start*
2006	70 million yen	Book publishing
2007	150 million yen	Various contributions, media publishing
2008	110 million yen	Book publishing, ditto
2009	80 million yen	Book publishing, ditto
2010	125 million yen	Book publishing, ditto

Fig. 37.2 Annual fluctuations in my contributions toward generating consulting-project orders

Source: Author.

In conclusion, by treating information and systems in the consulting field, which is a field that leverages expertise, and by carrying out mediative profiteering through information transmissions, *NRI* was able to not only produce effects that were intangible but also those that left behind their traces in the form of substantial numerical values.

Bibliography

1. Accenture (Toru Murayama, Koji Mitani, CRM Group/Strategy Group), *CRM — Customers are Just There,* Toyo Keizai Shinposha, 2001.
2. Nomura Research Institute, Enterprise Currencies Project Team, *Corporate Currencies of 2010 — The Point Economy of the Googlezon Age,* Toyo Keizai Shinposha, 2006.

38

Rapid Expansion of Partnerships Through Participating in Common Points Programs

Hiromichi Yasuoka

Partnerships and Issuance Volumes Made Possible Through Point Programs

Sales promotion costs and the advertising and general publicity expenses of companies amount to the magnitude of around 13 trillion yen and around 6 trillion yen respectively every year. However, advertising expenses in particular these days have been seeing a continuous decline. A major factor behind this trend is the fact that many companies, amid a long lasting recession, are controlling expenses whose returns on investment cannot be understood. For this reason, what is attracting attention now is the points and coupon programs.

The specific reason why the issuance of points has increased is because points programs directly return benefits and privileges to end users, unlike advertisements placed in various media. In other words, it is easy to verify whether the granting of points have had any effect by varying the rates of points granted by every product and by linking to Point of Sales (POS) systems. In addition, for members of points programs, direct approaches have become easier by associating them with ID numbers (special membership numbers). Furthermore, another reason is the fact that networks between retail shops and headquarters have become easier to build, thanks to the spread of broadband.

End users get to enjoy the pleasure of saving points and using them. These accumulated points are used for companies that grant the points (and other companies that have tie-ups with these companies). This differs from price discounts, which offer no means for a company to know where the savings brought about by the discounts come from.

The major objectives of these points programs are:

1. Customer enclosure;
2. Cultivation of loyalty;
3. New customer acquisition;
4. Reciprocal customer transfer.[1]

In other words, as shown in Fig. 38.1, the points program, in addition to offering direct returns to members, not only encourages customers to repeat their patronage for the company offering them the returns, but also makes it possible to transfer customers among firms enjoying tie-ups with the company originally issuing the points.

As seen above, the trend for companies to utilize points that offer direct returns to customers and to form tie-ups with other firms (or carry out CRM through multiple companies) is becoming more mainstream than for a single company to carry out points programs on its own. Incidentally, Fig. 38.2 shows point-partnership programs that also include those that make use of electronic money (programs that entail the exchanging, offering and use of points among partners). Strictly speaking, this diagram reflects the main parts. In actuality, the configuration it describes is as

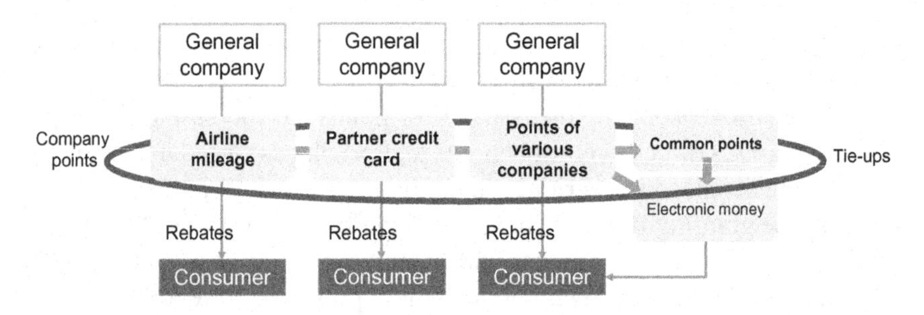

Fig. 38.1 Diagram showing direct returns and business tie-ups

Source: Author.

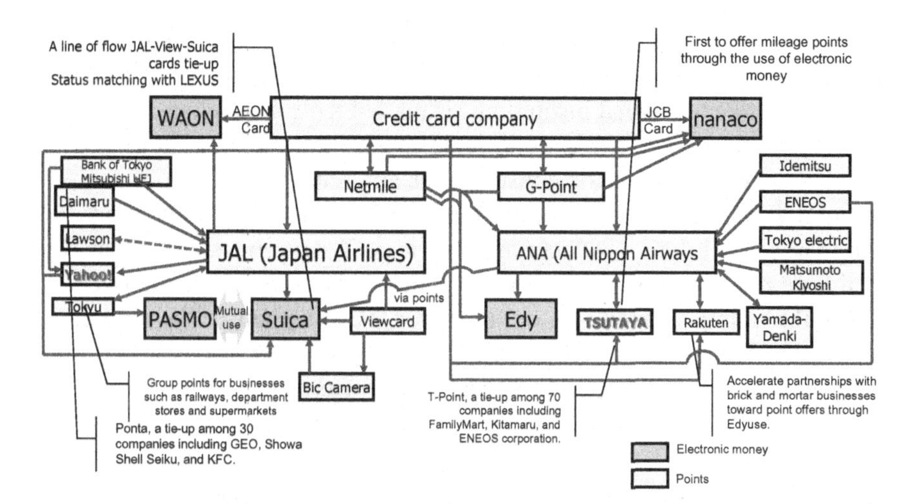

Fig. 38.2 Major points programs and partnerships with electronic money businesses
Source: Author.

complex as a circuit diagram, so for further details please refer to relevant magazine reports.

The points program is a marketing tool that has been around for a long time. The current electronic Japanese point-card system is said to have been started by *Yodobashi Camera* in 1987, but similar schemes had already existed long before that. For example, schemes such as the bell-mark scheme that entailed assigning bell-marks to various products and those that entailed pressing stamps of approval on paper were all schemes that involved offering various types of points.

Currently, in the case of Japan, not only private companies, but various other types of organizations are issuing points, ranging from public companies to governmental authorities. Incidentally, the points programs enjoying the most popularity among consumers are mainly found in the daily-life retail sector (supermarkets, drug stores, consumer electronics retail stores, cell-phone shops and gas stations). In other words, the points programs deployed by businesses in this area have the most consumers who consciously accumulate points.[2]

Plenty of companies make use of these points programs and the total amount of points issued in 2009, as estimated by Nomura Research Institute, exceeded 1 trillion yen at least when adding government-issued consumer electronics eco-points. While lately there has been a move toward narrowing down the rates of granting points by industries that have issued too many points (credit card companies and mass retailers of consumer electronics), the range of companies offering points in one way or another, from public companies to educational institutions, is undoubtedly expanding.

Incidentally, this 1 trillion yen scale means that the amount of points granted was for a level of consumption amounting to 100 trillion yen, assuming that an average of around 1% of points were granted for purchases of various products and services. When considering that the real final consumption expenditure of households was 227 trillion yen in 2009, the points granted turn out to be for less than half of the total consumption. Therefore, it can be said that there is scope for issuing more than half the amount of points that had been offered.

Rapid Partnerships that can be Obtained by Participating in Common Points Programs

Among points programs, the ones that are attracting the most attention these days are the following programs: the *T Point* program, the *Ponta* program, the *Rakuten Super Point* program, and *JAL* and *ANA*'s programs that offer mileage points. The number of members in these common points programs as of May 2011 was 37 million for the *T Point* program, 32 million for the *Ponta* program, more than 60 million for the *Rakuten Super Point* program, and 20 million each for *JAL*'s and *ANA*'s programs. Many users can be reciprocally transferred among companies participating in common points programs.

These common points programs have many affiliated stores and many places where users can accumulate points. Users are also increasingly converting the points they accumulate from other companies to common points. While a time lag occurs in the course of normal point conversions, schemes that help attract users more by allowing them to make immediate conversions have appeared. A case in point is the scheme that allows users to convert from *JCB* points to *Yahoo! JAPAN* points. Such a scheme is also capable of directly attracting users by not only allowing them to exchange common points but also by offering them these points from the beginning. When such immediate point conversions and point offers increase, it becomes more possible to thicken the lines in the tie-up diagram and further accelerate cooperation.

In brief, when participating in these common points programs, companies can carry out business tie-ups that they wouldn't have been able to carry out on their own otherwise. By doing so, companies can increase their customer traffic at once. For example, with the *T Point* program, tie-ups among 70 some companies have already been established and when a company in this program seeks to secure an average amount of members, it will be able to carry out partnerships with that many companies. In the case of the *Ponta* program, cooperation exists among more than 30 companies, making it possible for a company within the program to instantly have business tie-ups with just as many companies.

An employee of one company has privately expressed that the company's customer traffic clearly saw a boost in the first year of its participation in particular. For instance, when customers considered whether to visit the store of a participating company or whether to visit a store of a non-participating company when the distances to both of these stores weren't that different, the customers ended up visiting the store of the participating company. Consequently, when customer traffic increases, inevitably sales also increase. In this way, participating in a common points program can lead to the attainment of a super effect.

However, the companies in these programs are those that require just an average number of members (end users) and for this reason most of them are not carrying out substantially effective marketing measures. The effects for companies already carrying them out are probably not dramatic. In addition, companies that are carrying out these measures tend to mostly be leading companies in their sectors and since their measures are considered to be focused on their own initiatives, their level of participation in common points programs has been low so far. Therefore, the effect is unknown.

However, as explained above, unlike industry leaders, it is not favorable for a single company to offer points programs on their own. However for companies considering offering these programs from now on, participating in a common points program is worth their consideration.

Bibliography

1. Nomura Research Institute, Enterprise Currencies Project Team, *Enterprise Currencies Marketing — Next Generation, "Recommendations on using points and electronic money,"* Toyo Keizai Shinposha, 2008.
2. Nomura Research Institute, Electronic Payment Project Team, *Electronic Payment Business — A service that goes beyond banks appears,* Nikkei BP, 2010.

Section 2: Cost Reduction

39

Reduction in Cleaning Costs by Approximately 60%

Tetsuro Saisho

A Conceptual Shift

A case saw a substantial cost reduction by changing the concept of a conventional business model that was taken for granted in its industry. This case concerns the recycling of cartridge filters for dust collection in laser cutters used in the field of assembly processing.

The laser cutter is a machine tool widely used in the field of assembly processing. Specifically, it is applied in operations for cutting iron or stainless steel materials. Laser cutters vary in kind by the type of laser used and the shape and size of the material to be cut.

At any rate, regardless of the type of laser cutter used, to achieve its stable and continuous operation, maintain a comfortable workplace environment in material processing sites, and ensure the health of the workers, the careful maintenance of laser cutters (including cleaning) and the regular exchanging of filters (including washing) are absolutely necessary.

Using laser cutters produces industrial dust, which is comprised of particles of metallic powder generated in the course of processing materials. These particles have an extremely large and toxic impact on the environment and on the human body. From around six months to a year, metallic powder particles begin to cling to the filters built into the laser cutters and form blockages, negatively impacting their capability to collect dust. Since the filters would cease to function adequately, the purchasing of brand new filters had been deemed a necessary part of the operational process.

In other words, from an operational perspective espoused by the conventional business model, where filters of the laser cutters with blockages caused by metallic powder particles were concerned, it was natural to consider them as disposable goods; it was natural to replace used filters with new ones.

The line of thinking (until now) held that replacing filters with brand new ones was the common and obvious response to take in terms of both cost and operational efficiency; to wash used filters and reuse them apparently appeared to be a labor- and time-intensive process. It is also costly since doing so apparently did not restore the filters' adequate performance and function.

Amid such a situation, *Gunkyo Factory Inc.* (Takasaki-shi, Gunma, President Noboru Toyama), a company producing nozzles for laser cutters, has established a filter washing-related technology that facilitates filter cleansing and reuse and maintains its performance and function at levels comparable to those of a brand new filter. Consequently, by foregoing the purchase of brand new filters and using them, the company has been able to reduce their filter cost by approximately 60%.[1] In effect, with this technology, the company has been able to shift away from the practice of disposing used filters to the practice of washing and recycling them. Consequently, the company achieved a 60% cost reduction.

Toward the Construction of a New Business Model with the 3R Concept

In fields that require assembly such as automobile parts processing, precision sheet metal processing and machining, to process metallic substances such as iron, stainless steel, carbide, steel, aluminum and titanium, workshops often make use of laser cutters, in which case, metallic powder particles is produced as industrial waste.

Generally to date, filters were replaced by brand new ones since their dust collection function ceased to work due to drops in pressure caused by blockages that formed after around six months to a year and due to a drop in the flow rate. In the exceptional case of washing filters, depending on the kind of metallic powder found in them, hydrochloric acid and alkaline-based industrial washing agents were used.

In addition, when companies using laser cutters carry out filter washing, they blow air into the filters to blow off dust and then carry out washing with high-pressure water. However it is extremely difficult to neatly remove the blockages by this washing process alone. This is because the hard carbon and metallic powder particles, which are industrial wastes, cannot be readily removed even with cleaning agents such as oils or industrial detergents, not to mention washing by blowing air and washing with water.

Moreover, if industrial waste remains (even in small amounts) when washing filters with water, more often than not, the rusting and adhering of iron content occurs. Consequently, the recovery rate of filters declines with each washing, shortening the service life of the filters which therefore increases the frequency of replacements.

Consequently, *Gunkyo Factory Inc.* developed a proprietary filter-washing technology that entails soaking used laser-cutter filters in special detergents and then manually applying a high-pressure jet of water and a rinsing agent repeatedly.

With this technology, the company realized the merits of the 3Rs — Recycle, Reuse, and Reduce: while filters were originally assumed to be disposable, the technology allowed for it to be recycled; the technology facilitated washing for reuse of the filters; the technology helped to prevent the substantial discharge of waste.

As shown in Fig. 39.1, by washing used filters to reuse them again without performing any modification, or in other words, by washing and

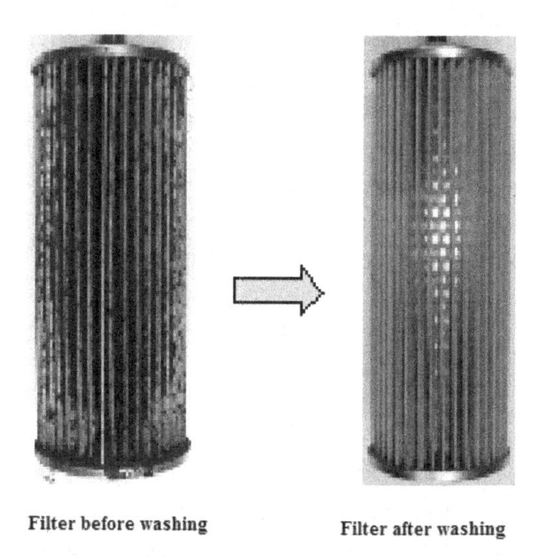

Fig. 39.1 Cartridge-filter cleansing

Source: Author.

reusing filters that had been treated as disposable, the company was able to realize a drastic reduction in costs related to filter replacements (running costs and disposal costs).

Additionally, the technology has helped to reduce the burden on the environment by facilitating energy savings made possible through maintaining filter efficiency and by facilitating the conservation of resources through a reduction in waste disposal.

Mission to Uphold Public Responsibilities

Many filters used in laser cutters are pleated and upon repeated use, industrial waste such as particles of metallic powders, degraded oil, and carbon begin to adhere to the elemental parts that constitute the filters, causing blockages.

Therefore what becomes key in the cleansing process is to improve the performance of the filters after washing them by removing residual substances such as grinding fluids and abrasives, organic dirt, and the extraneous industrial waste that adheres to filters in the process of cutting metallic substances. The air permeability of filters after washing them can be revived to approximately 95% of a brand-new filter's air permeability, and if they are washed repeatedly they will be able to be used on a semi-permanent basis.

However, the task of cleaning dirty filters is both labor- and time-intensive and is dirty and reqires manual labor, workers tend to dislike and shun this task, apprehensive about the difficulty they will face with this task. In addition, there is also the work-environment-related issue in which workers might breathe in poor-quality air while cleaning filters and becoming unhealthy.

For this reason, if a company's cleaning operation is outsourced to a third party that specializes in this type of work, its business will sufficiently benefit (even it only a small level of the cleaning work is outsourced), since the specialist third party will handle the task in a concentrated fashion.

Gunkyo Factory Inc. has entered into a new business area that specializes in cleaning cartridge filters for dust collection, targeting a clients who purchase their laser cutters. They are also extending their business relevant to this new line.[2] For instance, companies making use of *Gunkyo Factory*'s cleaning service can send their dirty filters to *Gunkyo Factory*'s workshop and expect to have washed filters returned to them within approximately two weeks.

A commonly used barrel-shaped standard cartridge filter for dust collection measures 16.5 cm in diameter and 54 cm in height and costs 5,900 yen. Replacing a brand new model one time will cost 16,500 yen, a sum that covers the purchase cost and the disposal cost. For this reason, more than 60% of such expenditure can be saved.

The effective utilization of limited natural resources is a large challenge that companies of the 21st century have to face. The recycling of cartridge filters for dust collection in laser cutters will not only reduce cost but also reduce the output of industrial waste and lead to the improvement of the work environment. Therefore, it can be considered to be a means for companies to demonstrate their mission to be publicly responsible for the future.

Bibliography

1. Nihon Keizai Shimbun, *Filter Cleaning Business,* October 8, 2010.
2. Gunkyo Factory Inc. website. Available at http://www.gunkyo.co.jp/ (20120930)

40

A 40% Cost Reduction in Transporting China-Bound Cargo

Tetsuro Saisho

From a Large-Scale Service to a Small-Scale Service

The case covered in this section examines how a drastic cost reduction was achieved after transforming a service from a traditionally large-scale operation to a small-scale operation. To be more specific, this case is about how a company was able to offer a service at prices lower than the norm by specializing in small lot cargos for routes between Japan and various Asian countries and regions and by offering regular transport services.

In the case of China, with its low cost and abundant manpower, the country to date has been functioning as the world's factory for manufacturing enterprises of many advanced nations, including those of Japan. However with the surging demand for wage hikes from workers and the high likelihood of the re-evaluation of the yuan against the US dollar, many manufacturing firms whose operations are labor-intensive are shifting their production bases away to other Asian countries where the cost of labor is low, such as Thailand, Malaysia, Bangladesh, India, Indonesia and Vietnam.

Impacted by such a shift of production bases of many manufacturing industries of various nations, Japan's logistics companies are reinforcing their transportation business that connects Japan to various Asian countries and regions.

Amid such a situation, *Mitsui-Soko Co., Ltd.* (Minato-ku, Tokyo, President Kazuo Tamura), which runs a warehousing business, harbor transport business, domestic forwarding business, international forwarding agency, and a real estate lease business, is cultivating new clients by developing and offering a low cost service that focuses on small lot cargos.

For its first Asia-oriented service, *Mitsui-Soko* has reduced the price for transporting goods to China by approximately 40%. To achieve this, the company has put together a sea-transport service provided by various shipping companies with services linking Japan to Shanghai, while securing container transportation through a shuttle flight service that operates five times a week.

The containers used by the company are standard frame-shaped boxes and are made with iron steel or aluminum. Cargos are loaded into them and transported to foreign destinations *via* aircraft or ship. In the field of containerized transportation, for most of the cases involving corporate customers until now, transportation was usually carried out by reserving one container per client.

However *Mitsui-Soko* adapted the seaborne containers (which it charters to meet its specific objectives) to accommodate multiple clients, accepting small quantities of cargos from each of them. In effect, the company made the container function like a communal bus in which many riders ride together. With the availability of this service, even medium- and small-sized businesses became capable of quickly transporting their products produced in China to Japan *via* Shanghai in as many quantities as necessary.

Promoting the Air Forwarding Business (International Cargo Transportation Business)

The Air Forwarding Business (AFB) is a freight forwarding business. It offers the most suitable means of transportation for shippers by handling (through consignment) networks of medium to long distance regular flights, chartered transportation, general freight traffic, and fixed temperature freight traffic. Generally speaking, the AFB business specializes in consolidating cargos to be conveyed *via* sea or air freight.

The consolidation business entails collecting cargo from various clients and consolidating them into one full load to be shipped in one container to various overseas destinations by shipping companies or airlines entrusted to transport the goods. *Mitsui-Soko* has been able to realize ideal load efficiencies by combining large, medium and small lots of cargos that had been traditionally transported in containers *via* chartered flights. With such efficiencies, the company is able to offer a distribution service at a bargain-level shipping cost that is approximately 40% cheaper than before.

Since the cost of transporting goods to China is usually determined by the size of the container, the volume and weight of cargo are generally insignificant factors. In other words, when the load efficiency is poor, there is considerable waste involved in the transportation. However this is where the consolidation business helps, namely by eliminating such waste and realizing reductions in transportation costs.

Now we shall explore transporting to Japan various goods produced in factories located in China, such as electronic components, automobile parts, devices, information and communication mechanisms, nonelectrical machinery, metalware, fiber, apparel, and food products. Up to now, many cases of transporting cargos from foreign countries involved using one container for one client and when the load was small, containers ended up having vacancies, making their turnover rate inefficient.

Additionally, in the consolidation business, when loading cargos from multiple clients into one container, there used to be occasions when shipping would become more expensive because of an increase in workload to accommodate some cargos that required different or special handling. Alternatively, there would be times when a single cargo among the consolidated mixed cargo load could not be delivered since it needed to be dispatched immediately.

Mitsui-Soko's new service made it possible to carry out efficient consolidation, allowing small quantities of cargos sent from multiple clients to be loaded into one seafaring container. Consequently, it became possible for the company to offer their services at prices ranging from 8,000 yen to 6,800 yen, providing services that are approximately 40% cheaper than what they had been charging previously. For the future, the company plans to expand its business network from China to regions throughout Asia to handle 500,000 tons of cargos per year and attain a revenue of 2 billion yen three years from now.[2]

Company name	Establishment	Summary	Features
Nippon Express, Nishi-Nippon Railroad, Kintentsu Express, Yusen Logistics	January 2011	Established a joint charter for deploying a cargo aircraft to service the Haneda-Hong Kong route and then entered into ontracts with Vantec, Hankyu Hanshin Express, and Yamato Global Logistics Japan	Concluded a split charter contract (direct charters by multiple shippers) that entails the use of freighters in Hong Kong (cargo flights)
Hitachi Transport System, Kintetsu Express	July 2011	Established partnerships with large-scale, infrastructure-related transport facilities located in Asia and the Middle East.	In accordance with the appropriate business strategy and area strategy, the companies collaborated and built highly advanced service setups and optimized their global supply chain
Yamato Holdings	January 2010	Entered into the express delivery business in Shanghai and Singapore and has plans to make the same market entries in Hong Kong in February 2011, and in Malaysia in September 2011	In Asia, economic growth has led to improving living standards, which in turn has helped to escalate consumer spending and prepare an environment suitable for the use of high-quality distribution services
Nippon Express	November 2009	Maintained truck-transportation network (SS7000) that links the distance of approximately 7,000 km spanning the Shanghai, China and Singapore route	Built a supply chain within the Asian region, which has turned into the world's production plant today
Hitachi Transport System	April 2011	Purchased the Thai Distribution company, Eternity Grand Logistics Public Company (ETG)	Expansion and reinforcement of the company's operations in the Indochina region with a focus on Thailand, where demand for distribution services is anticipated to increase with economic development

Fig. 40.1 Development of major Asia-related physical distribution businesses

Source: Created by making additions and amendments to the data that appeared in the "Nihon Keizai Shimbun" on May 23, 2011.

Expansion of the Scope of the Logistics Business

As shown in Fig. 40.1, impacted by the expansion of the economies of Asian countries, major domestic logistics companies are reinforcing their business in Asia. These companies are *Nippon Express Co., Ltd.* (Minato-ku, Tokyo, President Kenji Watanabe), *YAMATO HOLDINGS CO., LTD.* (Chuo-ku, Tokyo, President Makoto Kigawa), *Hitachi Transport System, Ltd.* (Koto-ku, Tokyo, President Takao Suzuki), and *Kintetsu World Express, Inc.* (Chiyoda-ku, Tokyo, President Satoshi Ishizaki).

Mitsui-Soko until now had not been supporting large-scale distribution, but by developing its consolidation business and by also supporting small lot distribution, they are stirring demand from mainly medium- and smallsized enterprises whose main business lines are fashion products and daily necessities.

Mitsui-Soko manages the workflow leading up to the loading of cargos, which ranges from documentation and export clearance to ground transportation. In addition, the company also provides services for delivering cargos to consignees after import clearance is handled by *Mitsui Soko*'s overseas branches and agencies who take over the mixed cargo loads from shipping companies once they arrive at their respective ports of destination.

With the surge in recent years of companies seeking price reductions in transportation charges along with the shifting of production bases to Asian destinations, *Mitsui-Soko* is planning to offer cross transport services in addition to services for cargos departing from Japan *via* sea. These services, will carry out arrangements for international transportation from *Mitsui-Soko*'s Japanese office. For example, such arrangements would be carried out to comply to a request made by a Japanese corporate client for having its cargo shipped *via* ocean transport directly to Shanghai from Ho Chi Minh without traveling *via* Japan, using the company's consolidation operations.

Incidentally, in the event of transporting to Japan products manufactured in factories located in China, for most of the cases, the products used to be stocked in warehouses in Japan. Since management expenses became more expensive in such cases, *Mitsui-Soko* also started a service for storing the products locally upon purchasing those products from the client company in China.

In addition, the company established the trading company *Mitsui-Soko Air Cargo, Co., Ltd.*, within the month. A 100% subsidiary, this company reduces transportation costs for companies by transporting purchased goods to Japan at any time in accordance with the client's needs. Aiming to expand and strengthen the international cargo transportation business further, *Mitsui-Soko Air Cargo* was established in March 2011 by acquiring all of the shares of *JTB Air Cargo*, an international cargo transportation business that is a wholly owned subsidiary of *JTB Corporation*.

This service transfers distribution processing services that were being carried out in Japan, such as storage, inspection and gauge examination, to China, reducing in total physical distribution cost and shortening the lead times through ocean transport services that depart from China almost every day. Furthermore, the company also offers other services that meet the needs of various clients through platforms and businesses that support mail order sales in China and temporary acquisitions to help improve cash flow for operations.

Bibliography

1. Nihon Keizai Shimbun, *Shipping cargo to China now 40% cheaper*, May 23, 2011.
2. Mitsui-Soko website. Available at http://www.mitsui-soko.co.jp/ (20120930)

41

Suica's Role in Generating Revenue and Reducing Operational Costs of Vending Machines

Hiromichi Yasuoka

Suica as Integrated Circuit Railway Ticket and Electronic Money

Suica's full-scale deployment as an integrated circuit (IC) railway ticket began on November 18, 2001. Initially this IC integration was implemented to rectify the numerous problems that arose in the use of the magnetic ticket gates, such as the trouble of tickets becoming stuck in the machines. However other purposes for its adoption have been raised as shown in Fig. 41.1.[1]

With the objectives of implementing the IC ticket, in March 2004, the company made it clear that *Suica* would have another use when it made a trademark registration for *Suica* to be used as electronic money (a prepaid card) that can be used for purchases made at distribution stores and from vending machines found inside stations and also those on the streets. Although *Suica*'s transactions involve relatively smaller sums compared to the sums processed by electronic money cards of other companies, since it is necessarily charged and used as IC railway tickets, the usage rate of *Suica* is high. Fig. 41.2 shows the major electronic money cards circulating in Japan. The railway-affiliated *Suica* is seeing a horizontal expansion as a foundational system for IC railway ticket systems and electronic money systems for Kansai's Japan Railway (JR)-affiliated

-Service upgrade	Handy to use and replaceable if lost
-System change	Prevention of ticket jams
-Cost reduction	Reduction in maintenance cost with automatic reset
-Security upgrade	Prevention of forged cards and abuse (cheating on the fare)
-New business	The possibility of developing new services and businesses

Fig. 41.1 Major purposes for introducing the IC ticket

Types	Independent	Railway		Retail	
Service name	*Edy*	*Suica*	*PASMO*	*nanaco*	*WAON*
Operation type	Rakuten edy	East Japan Railway Company	PASMO (Kanto Private Railway)	Seven Card Service	AEON RETAIL/AEON BANK
Service launch	2001/1	2004/3	2007/3	2007/4	2007/4
Number of cards issued	Approximately 60 million	Approximately 32 million	Approximately 17 million	Approximately 13 million	Approximately 17 million
Number of affiliated stores	Approximately 250,000	Approximately 120,000		Approximately 80,000	Approximately 100,000

Fig. 41.2 Comparison of major IC-based electronic money cards (prepaid types)

Source: Based on releases from each company.

ICOCA, the private railway-affiliated *PiTaPa*, Tokai's JR-affiliated *TOICA*, and many other railways across the country, including those of Hokkaido and Kyushu, not to mention *PASMO*, which is affiliated with Kanto's private railways. In this way, it is helping to reduce maintenance costs (achieve cost reductions) and prevent fare cheats.

The Ripple Effect of Introducing *Suica*

After *Suica*'s introduction, it is known that *Suica* was being considered to function solely as an IC ticket platform while *Edy* was being considered to function solely as an electronic money platform. Both of them were to be incorporated within the same IC card (in the manner of containing different wallets).[2] However in the end, *Suica* also assumed the function of electronic money, creating a disorderly situation where every company began to issue electronic monies.

The commission rates for affiliated stores charged by the companies specializing in the issuing of electronic money are not as large as the rates charged by credit card companies, since electronic money itself deals in small sums and moreover it is a form of prepayment so the risk of failing to collect credit obligations is minimal. Consequently, the revenue from

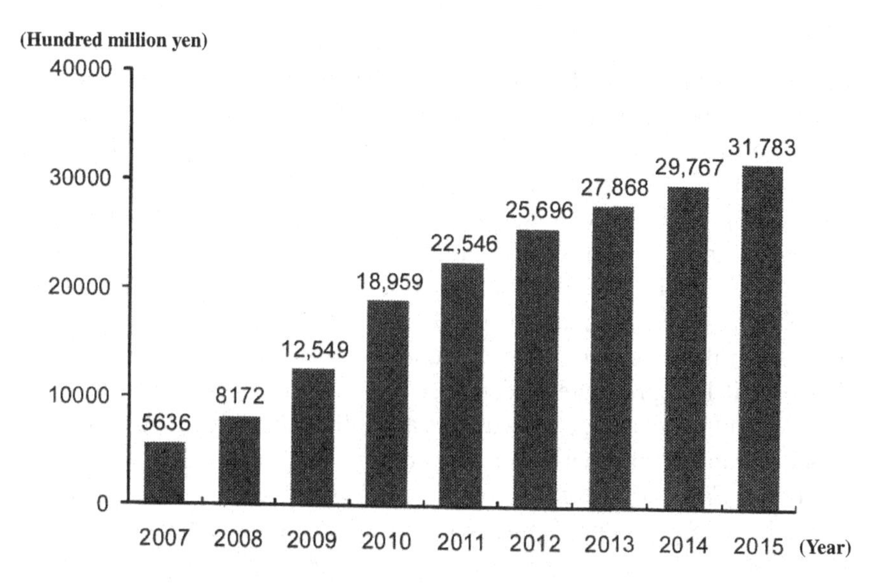

Fig. 41.3 Contactless IC electronic money

Source: Nomura Research Institute, *What's Going to Happen in the Information and Communications Markets From Now On — IT Shijyo Navigator (IT Market Navigator), 2011 edition".*

electronic money through commissions is not substantial either. Due to this low level of revenue, the reality is that companies haven't been able to achieve any return on their initial investments for electronic money scanning terminals. This is attributed to the fact that each company had to make these investments on their own as a result of the impact of the flooding of issuing companies into the market in a disorderly fashion.

Nonetheless electronic money has been impacting Japan's consumption in a large way. As shown in Fig 41.3, compared to the market size in 2007, which was supposedly "the first year of electronic money" according to the Nikkei Marketing Journal when *PASMO*, *nanaco*, and *WAON* were all launched, in 2010 the market was more than three times the size then and is forecasted to grow by more than three trillion yen (five times more than the size in 2007) from now on.[3] Of course, compared to the time before the first year, the growth has been dozens-fold (considering that it was originally zero). Amid stagnant consumption stemming from a sustained recessionary period, the extent of this growth indicates a superlative effect on consumption.

An exemplary case in point is the vending machine. *JR East Water Business Co., Ltd.* (Shibuya-ku, Tokyo, President Osamu Tamura), which handles JR East Japan Railway Group's beverage business, anticipates the sales turnover of its vending machines to be around 26 billion yen for the March 2011 period. Compared to the turnover of 18,700 million yen registered in the March 2007 period, the growth is nearly superlative. Until recently, there had been no changes in the total number of vending machines so the increase in revenue must have been achieved by the sales growth attained by the vending machines on a per-unit basis. This is largely attributable to the introduction of *Suica*-supported vending machines and their automated capability to acquire customer data, perform analysis on them, and translate that analysis to merchandise assortments. However, it is appropriate to consider that the effects derived from such analyses and merchandise assortments are still forthcoming, and that at the outset, it can be said that it was the convenience made possible through supporting *Suica* that largely contributed to an increase in sales. In other words, it was *Suica*'s speed-purchase maximization effect. Incidentally, for the period of July 2011 *Suica*'s payments rate was 45% and the number of *Suica*-supported vending machines was approximately 6,800. Amid the low price trend of the times, and for the vending machine industry that remains in a stagnant state, *Suica* is proving to be a catalyst.

In addition, it is believed that *Suica* alone was responsible for an increase in revenue by more than 10 billion yen just on the basis of its capability to carry out prevention against fare cheats after the IC ticket was introduced.[1] Incidentally, even if this increase can be strictly accounted for by other factors, approximately 3,500 million yen cannot, so in a strict sense, at least 3,500 million yen can be said to be the result of the *Suica* effect.

There were several factors behind promoting electronic money besides convenience. For example there was the points program. According to the "Questionnaire Survey on Electronic money" (June 2010 Internet research) carried out by Nomura Research Institute, among the reasons to use electronic money, "I can receive points and discount services" was the No. 1 reason at 41%. This is indicative of the fact that electronic money, a payments system that promotes its own use through its convenience, was also responsible for raising awareness of points programs, owing to the fact that points could be obtained through its use. Additionally, as shown in Fig. 41.4, the saved points were often converted to electronic money. And so these were the ways in which electronic money and points programs worked together to realize mutual expansion.

The above-mentioned digital monies had wielded a great influence on Japan's consumption habits. While railway-affiliated digital monies have

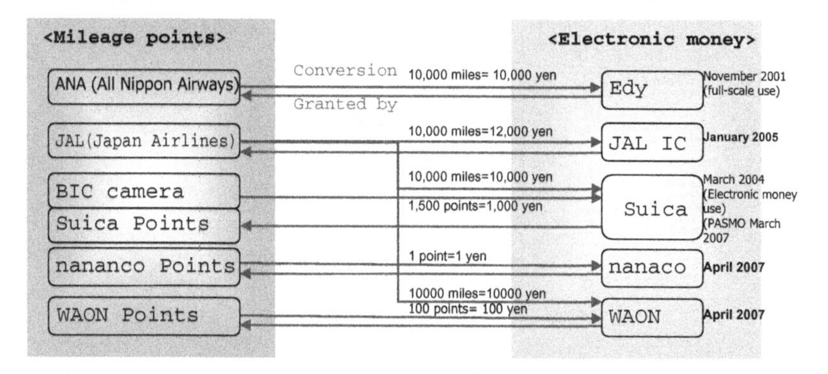

Fig. 41.4 The recycling economic system of electronic money and mileage points programs
Source: Author.

been available in various other countries, in Japan you can find the use of even distribution-related and independent electronic monies, making Japan unparalleled by any other nation in terms of the expansion of the electronic money market. In other words Japan is a major electronic money power.

By skillfully leveraging the conveniences and other advantages of these digital monies, their use can be further promoted and their usage track record (big data) can be made to grow vastly. While there are still only a few companies capable of skillfully applying these aspects of electronic money, the effects of analyzing the vast repository of lifestyle data (life-log) is thought to give rise to a further super effect.

Bibliography

1. Akio Shiibashi, *Suica will change the world — The lifestyle revolution triggered by East Japan Railway,* Tokyo Shimbun Shuppansha, 2008.
2. Yasunori Tateishi, *The truth behind FeliCa — The reason why Sony succeeded in its technological development but failed in running it as a business,* Soshisha Publishing, 2010.
3. Nomura Research Institute, *What's Going to Happen in the Information and Communications Markets From Now On — IT Shijyo Navigator (IT Market Navigator) 2011 edition,* Toyo Keizai Shinposha, 2010.
4. Nikkei BP Marketing, *Nikkei information strategy,* April 2011 issue.

42

Business Opportunities in Ultracompact Orbiter Development

Akira Ishikawa

Establishment of the Low-Priced Space Utilization Process

When speaking of new products and new businesses that are producing remarkable super effects, we cannot do without mentioning the effects the artificial satellite and its development and business operations have on the space aviation industry.

In 1990, after the American trade policy "*Super 301*" was put into effect, by agreement Japan had to make international bids for its satellites, even if they were to be used domestically for commercial purposes. However, since Japan had not been mass manufacturing satellites, its satellites were costly, making it unable to compete with the commercial satellites of European and American companies, which had been producing low-cost satellites through bulk production. For this reason, for example, for the successive satellites of "*Himawari No. 5*" Japan was put into the situation of having to purchase from a U.S. corporation the Multifunctional Transport Satellite (MITSAT).

To recover from such humiliation, the direction of Japan's artificial satellite development turned toward compact and ultracompact (or nano) sized satellites. Until several years ago, satellite development mainly focused on large-sized satellites that weighed over 500kg and never cost less than 50 billion yen. Such satellites included the Mobile Broadcasting

Satellite (MBSat) for movable bodies used by both Japan and Korea for communication and broadcast functions, weighing over 4,000 kg, and the Advanced Land Observing Satellite (ALOS), "*Daichi*". Certainly this was a major detriment toward commercialization, and what was worse was that the more Japan tried to improve its reliability, the price of exclusive parts tended to rise from several thousand times to tens of thousands times the price of generic parts.

For this reason, by the 1990s, the domestic content of the major parts of artificial satellites, which used to be 70%, became less than 30% with major companies pulling back.

However, what we should not overlook is the fact that a 10 cm^3, 1 kg-class nano satellite has actually been flown into space.

For example, ever since Tokyo University's so-called CubeSat sized satellite XI-IV (sai-four), which is shown in Fig. 42.1, and Tokyo Institute of Technology's CubeSat sized satellite CUTE-I satellite was successfully launched in June 2003, the number of miniaturized satellite productions and launches from mainly universities and town workshops has seen a gradual increase.

What is driving this rise is the objective to establish a speedy, space utilization process, and gaining a competitive edge in satellite development.

Fig. 42.1 CubeSat's Satellite XI-IV (Sai-four)

Source: Nakasuga Laboratory, Tokyo University, "Tokyo University CubeSat Project"

In effect, the aim is to realize a double-digit reduction in costs, including the cost of launching, that is around 1/500th of the previous cost so as to bring it down from an average of 50 billion yen to 100 million yen and shorten the time from order to delivery to around one and a half years and also aim to make it possible to carry out swift space demonstrations of new technologies and the development of new space utilizations. In addition, they are also examining issues related to acquisitions of communication frequencies and damage compensations, problems related to infrastructural arrangements and use to advance preparations for the commercialization of miniaturized and nano satellites.

Venture Business and Parts Research

In recent times, what has further strengthened this movement is the formation of the "Next-Generation Space System Technology Research Union," a collaboration between venture companies based in Yokohama that are developing components for artificial satellites and the researchers of Tokyo University. This union's target is to keep the cost of artificial nano satellites that can be used in natural resource explorations to around one-tenth of its current cost and make a foray into the global market.

What was fortunate was that the union was recognized by the Ministry of Education, Culture, Sports, Science and Technology as an enterprise eligible to be part of the "Advanced Research and Development Support Program" and was earmarked to receive a total budget of 4,100 million yen by 2013. This program was included in the Ministry's 2009 supplementary budget.

Presently there are five participating companies, including *Orbital Engineering*, a small business in charge of insulation specifications for satellites and *AXELSPACE Corporation*, a venture business that originated in Tokyo University. This group will be led by Professor Nakasuka Shinichi of Tokyo University's School of Engineering and will begin development with the cooperation of researchers from at least eight universities as well.

For the future, to overcome the issues of a vast development cost, constraints of the launching location, and timing constraints including adjustments to be made with concerned local fisheries, they are beginning to consider an air launch system for the artificial satellites, just like for the H2A rocket, Japan's major large-scale rocket. This would entail attaching a satellite-loaded rocket to an aircraft and launching it from a considerable height above the high seas. Consequently, the plan calls for separating the satellite and releasing it into orbit.

U.S. Companies Begin to Pursue

However Japan cannot afford to be negligent even in matters of a nano satellite launch such as this one.

For example in August 2009, the American company Interorbital Systems announced that it will launch a service called "TubeSat Personal Satellite Kit" from the fourth quarter of 2010, which is a low cost satellite service for individuals, costing only 8,000 dollars per day.

The finished satellite is slated to be launched into low orbit at an altitude of 310 km using the company's rocket *NEPTUNE30*. The control and management of the satellite after launch, however, is to be carried out by the purchaser of the kit through the use of an amateur radio band.

Bibliography

1. Intelligent Space Systems Laboratory (Nakasuka Laboratory), The University of Tokyo, *7,000 days have transpired since the completion of the XI Series,* Tokyo University CubeSat Project. Available at http://www.space.t.u-tokyo.ac.jp/cubesat/ (20110528)

Section 3: Sales Expansion

43

Rakuten's Sales Expansion by Entry into the Financial Business

Hiromichi Yasuoka

Increase in *Rakuten*'s Share of the Financial Business

Rakuten, Inc. (Shinagawa-ku Tokyo, President Hiroshi Mikitani), a company running the so-called *Rakuten* market in the shopping mall of electronic commerce, has lately grown through its financial business. Since 2003, it actively carried out mergers and acquisitions (M&As) and turned its online financial business unit (Internet Securities) into the second pillar of its group.

At the outset, as shown in Fig. 43.1, the company's finance business began with the purchase of the online securities firm, DLJ Direct SFG Securities (currently *Rakuten* Securities). The company then went on to purchase a credit card company, a consumer credit company, an online bank, and an electronic money company. Consequently the proceeds from its financial business already accounts for nearly half of the company's overall business now.

Since the company's own card naturally becomes involved with the company's own bank and electronic money as a means of payment for transactions that take place in areas other than their internet service business (*Rakuten* Market) and in even other areas than that, the impact is even greater.

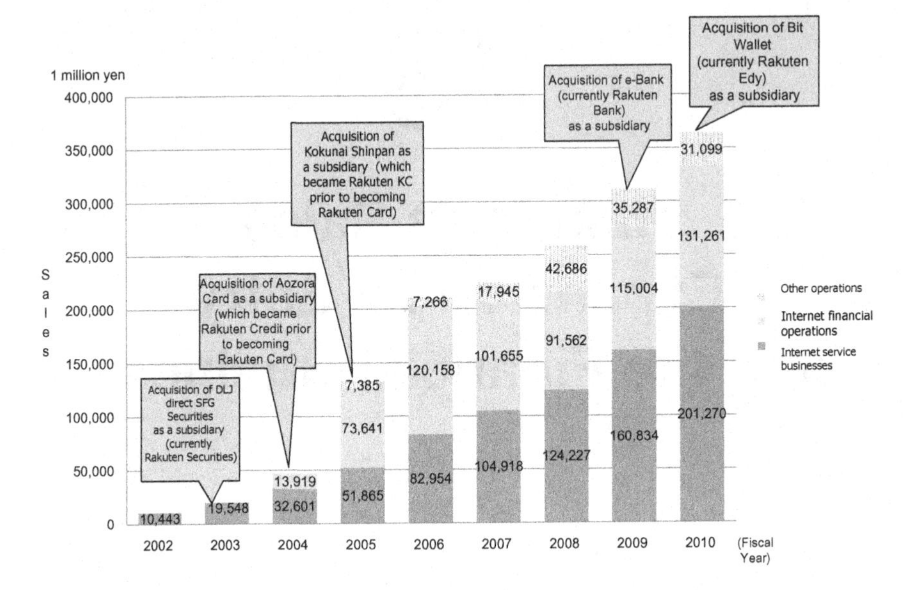

Fig. 43.1 Fluctuations in *Rakuten*'s sales and its financial business expansion ***via*** M&As

Note: Since the sales proceeds of the company's financial operations are unclear in the consolidated statement for 2003, aggregates are displayed.

Source: *Rakuten*'s statement of accounts.

The Synergy Between the *Rakuten* Market and its Financial Business

Since entering into the financial sector in 2003, the company's turnover has already become 20 times more than its turnover in 2010. So what factors lie behind such a continuous expansion?

Rakuten's core business is the *Rakuten* market. This is the key difference between *Rakuten* and another company that *Rakuten* was once compared to in the Internet bubble era: *Livedoor*. Although *Livedoor* repeatedly acquired securities-related financial businesses, there wasn't much synergy between its companies and *Livedoor* was evaluated to have unfairly raised the stock prices to carry out acquisitions through stock swaps.

Meanwhile, *Rakuten* has placed the *Rakuten* market as its core line. To guide users toward this business, the company makes use of web advertising and even makes good use of search engines (*Google*, *Yahoo*!) so that users end up arriving at the *Rakuten* Market even *via* them. Once there, users are invariably tailed with opportunities for making payments. For this reason, in addition to *Rakuten* Credit, which carries out loan services,

Rakuten went on to promptly purchase *Kokunai Shinpan* (currently *Rakuten* Card), which handles the most general-purpose credit card on the Internet.

If a user makes use of the *Rakuten* Card in the *Rakuten* Market, he or she can win a 1% *Rakuten* Super Point for using it in the market and in addition over 1% for using the card itself. This differs from the cards of other companies, which grant only 0.5%. In this way, the company accelerated the use of the *Rakuten* Market and the rise in the number of *Rakuten* Card subscribers. In addition, since *Rakuten* began granting *Rakuten* Super Points for making payments *via* means other than the *Rakuten* Card, the company was able to bring about a circle, motivating users to access the *Rakuten* Market for the purpose of using up those points as well.

In addition, the *Rakuten* Super Points valid for these payments have the merit of doubling in value when using several services called *Ponkan Services* offered by the *Rakuten* group. In this way, the company was able to raise the unit value of payments made by existing customers of the group.

Furthermore, adding to the company's credit card and points programs business, the company obtained *e-Bank* (currently *Rakuten* Bank), which is an Internet-based bank used for payments. In this way, it became possible for *Rakuten* to offer bank accounts that served as accounts from where credit card withdrawals could be made, making them a means of payments themselves while also turning them into accounts that could be used to cash out points. Alternatively, *Rakuten* had introduced a form of electronic money for online transactions, *Rakuten* Cash, from an early stage and with the enforcement of the "act regarding fund settlements" in April 2010 lifting the banks-only restriction for carrying out a cash transfer business, *Rakuten* entered into the cash transfer line as well. Consequently, it became possible for users to easily send cash to each other without having to go through a bank, raising the level of convenience for customers.

More recently, *Rakuten* acquired *Bit-Wallet* (currently *Rakuten* Edy), which issues the electronic money *Edy*, to attempt to draw in *Edy* users to the Internet from the brick and mortar world; in effect *Rakuten* wished to attract such users to its online destinations not only through providing a means of payment. Of course, the contributions from the company's professional baseball team *Tohoku Rakuten Golden Eagles* and the company's professional soccer team *Vissel Kobe* are also significant, owing to their wide name-recognition effect. However since the approximately

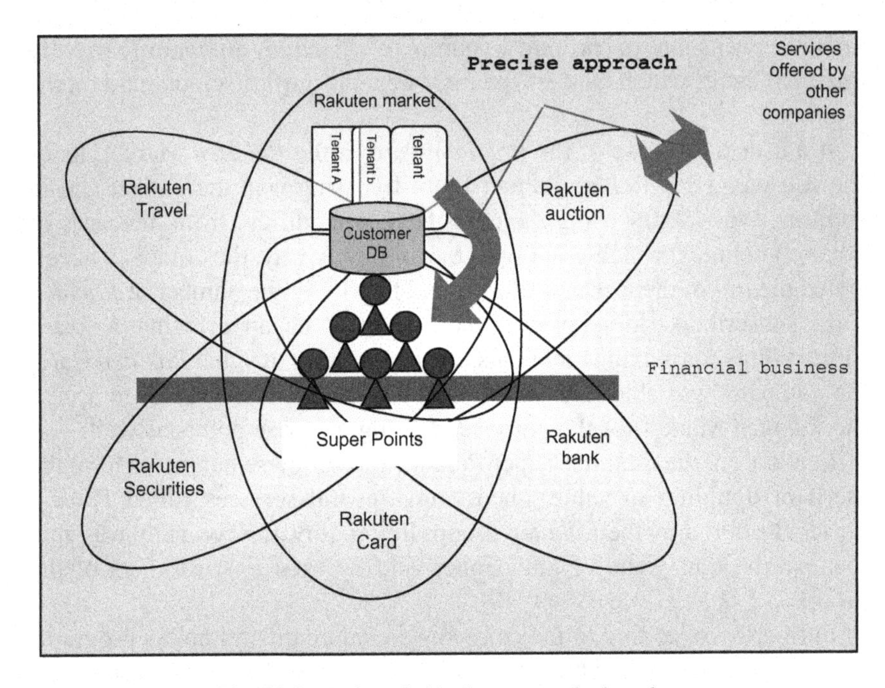

Fig. 43.2 *Rakuten*'s ideal group marketing plan

Source: Author.

60 million *Edy* users mostly make use of the electronic money at brick and mortar sites, the company's engagement with it should also be quite significant.

And so due to factors such as the above mentioned points regarding *Rakuten*'s financial business and the fact that *Rakuten* had developed an effective marketing strategy for its group of companies as shown in Fig. 43.2, the company was able to achieve a super effect of a more than twenty-fold increase in sales for its entire group.

The Future of *Rakuten*'s Financial Business

As explained above, *Rakuten* developed various financial business lines, making arrangements to provide a means of settling transactions used by users in all types of scenes, promoting user retention and facilitating connectivity in many ways. By doing so, *Rakuten* will be able to take their initiatives up to the level of customer enclosure and tie it to improving customer unit value in the future.

However, despite the fact that *Rakuten* had achieved a super effect, the company announced on June 2, 2011 that it will be restructuring its credit card line. In effect, this announcement had been an indication of the company's quick response to a change in its environment; namely the tightening of regulations for credit card services. The restructuring in particular involved breaking up *Rakuten KC*, the credit card business subsidiary and transferring its *Rakuten* Card business to *Rakuten* Credit while transferring *Rakuten* KC itself to the money-lending line, *J Trust*. In doing so, the company realized a concentration of management resources in its *Rakuten* Card business, a company for which high growth had been anticipated.

In addition, since *Rakuten* already has more than 60 million users within Japan, a growth achieved through expansions of each business will be ultimately limited. Therefore *Rakuten* is not only planning an overseas expansion of its presence in electric commerce (EC) but also an overseas expansion of its payment system from now on as well. In the future, the pros and cons of the success of these expansions will prove to be the group's touchstone to further growth.

44

CECIL McBEE's Product Development Model

Atsushi Tsujimoto

CECIL McBEE's Remarkable Sales Revenue

The fashion brand *CECIL McBEE**, run by *Japan imagination* [sic], Co. Ltd. (Shibuya-ku, Tokyo, President Hiroyuki Kojima), is continuously achieving superior sales. I would like to offer my commentary on this brand's unique product development and marketing styles.

Standing out in Harajuku, the fashion-business building "*Shibuya 109*" is a mecca for young fashion-conscious women. The area's average sale per 3.3 m^2 per month is 1,250,000 yen, which is more than three times the scale of department stores in Tokyo (as of 2010).[1] Among the establishments in *shibuya 109* that have superior sales records, *CECIL McBEE* has been consecutively No. 1 for the last 10 years and since the company is said to sell an average of seven times more than other stores (as of 2010), as a fashion brand *CECIL McBEE* boasts a sales revenue that exceeds the imagination by far.[2] The salespeople handling this brand are said to be truly in love with their own brand and apparently the desire they convey to buy the products of their own store is so strong that a shopper risks becoming flat out broke.[1]

In the store, it is obvious how deeply the customers are enamored with the salespeople there. Mr. Tatsuo Kimura, the Chairman of Japan imagination, the company running the brand, says, "If they don't like the brand we can't recommend it, and if it doesn't help make them look good, we can't recommend the brand again either". There truly are plenty of factors driving the sales of this brand. For example, one factor believed to have greatly contributed is the company's constant attention to store development — their consistent attempt to devise creative product displays and to work on the look of the store.[2] However in the end what has actively contributed to significant sales is "the act of observation" that supports the development of their original products and "the act of observation" made by the company's salespeople to decipher the consumer's tastes.

The Act of Observation to Support Product Development and the Salesperson's Act of Observation

Simply put, the act of observation supporting product development is the calm and patient "observation of the street". The director of the product development team says, "We watch what types of attitudes people project with the clothes they are wearing when they're walking down the streets. We observe at length the "tastes of the now" of people walking there.

Mr. Kimura makes it a point to mainly go to areas where his company's stores are located. "I want to know the discrepancy between the product and what the customer demands. There are the currents of the times to consider as well. Since it speeds past you, if you don't keep your eye on the ball you'll simply miss it..."¶

His comments indicate that he is concerned about doing whatever he can to avoid missing trends that could be described as "tastes of the moment" of general consumers for their attire, and whether these trends coincide with his company's brand to the fullest extent possible. Additionally he analyzes present trends that strictly evaluate the value of fashion from the "customer's perspective" in the following way.

> The choices young people make when they buy clothes are based on their youthful, personal sense of values. What's happened is a great transformation where customers now lead fashion trends by themselves. This happened because high school girls grasped the initiative to determine the course of products ... As long as CECIL McBEE continues to precisely grasp what the needs of those "gals" are, it won't disappear.
>
> Until then, fashion used to be just information granted from a high place with authority. For example, Paris, Milano, New York. The practice was to gratefully receive this information and make it break out during the next

season or one year later and then track it. Japanese consumers grew and matured and the leadership shifted to the customer.

— Mr. Kimura

Before the emergence of the societal transformation that saw the "transmission of fashion from consumers (or before fashion became grounded in the consumer's perspective), consumers used to attain contentment by subscribing to certain types of authoritative frameworks (authoritative information). However consumers began to gradually become faithful to their own tastes and inclinations. Since these tastes are based on the criteria of individual satisfaction and are ultimately idiosyncratic, compared to value criteria imported from foreign countries, current domestic fads change more quickly. Mr. Kimura had made that realization.

Next, let us examine the salesperson's act of observation. A salesperson attempts to communicate with a shopper in the way a friend communicates to another friend. His or her "act of observation" *vis-à-vis* the customer is indeed fine-tuned. During the conversation with the customer, the salesperson writes out on a receipt paper the "details of the conversation", "preferences", and "what the customer has". By repeating this task it is said that the salesperson can come to understand customer predilections. Consequently, he or she can start to recommend products that appeal to customer preferences. For the customer this will lead to the joy of recognizing that the salesperson appreciates his or her tastes and thus will encourage the customer to return again.

Some of the salespeople were originally customers of the brand themselves and therefore understand the tastes of the people visiting the store. The fact that there are salespeople whose tastes are similar to the tastes of the customers helps the company quickly understand the purchase inclinations of the targets. For a salesperson, the act of attempting to grasp a "mental imagery" of coordination, styles and tastes sought by consumers is a unique act to carry out.

The Nesting Situation of the "Observational Acts"

The results of the above-mentioned "act of observation to support product development" (A), help facilitate the success of "the salesperson's act of observation" (B), in effect putting them in a nesting relationship with each other. That is to say the mental image emerging from A's act of observation is found in the condition of being dissolved in the mental imagery formed through B's act of observation. By extension, these points can be summarized as follows.

1. Constantly continue to observe the "contemporary tastes" of potential targets (women aged 16 through 22). And then work on the development of original goods or promptly assess the prospects of goods that coordinate with existing products. (Observation Act A)
2. Constantly continue to observe the tastes and assess the mental imageries of customers who come to the store. And then ask what types of products among those handled and what kinds of product coordination can be recommended in a timely fashion. (Observation Act B)

The process encompassing 1 to 2 is the act of connecting the organization that supports the brand called *CECIL McBEE* to the "emotional world" of the consumers. The process of these two kinds of observational acts (the results of creative and deliberate thinking) piling on top of each other and becoming multilayered is believed to be maintaining the brand image and preserving consistency while functioning as a driving force that continuously fuels the company's super effective sales power.

Notes

*This is a fashion brand for "gals" with the concept "sexy and gorgeous," targeting females aged 16 through 22.

**At CECIL McBEE's "109" storefront, the pedestrian traffic is heavy and the display there is changed repeatedly during the day to project a fresh image of the store at all times.[3] Every three weeks the company invests in new products and updates its merchandise mix. Considering that other stores do so on average at a pace of once every two months, new clothes are put on display at CECIL McBEE's store at twice the speed.[2] "*FOREVER* 21," a fast fashion chain that became popular in the US, also changes its merchandise line frequently. This strategy works by inducing customers to think, "I must buy this product while it's still there". Retail companies that sell through stores often spur consumption behavior by adopting tactics such as varying visual sensations, projecting dynamism or a lively quality to impress passers-by or by activating the sensitivities of consumers visiting their stores.

¶Details on the format of customer communication applied at stores, comments made by Mr. Tatsuo Kimura, the chairman of *CECIL McBEE's* parent company *Japan imagination*, and comments from the director of the product project team are all drawn from Reference No. 2 shown below.

Bibliography

1. Yoshihiko Kadokawa, *The structure of "the profitable smile"—If you can open the hearts of customers, you can open their wallets too!*, Daiyamondosha, 2010.
2. TV Tokyo, *CECIL McBEE—Behind their miraculous sales,* Cumbria Palace (Ryu's Talking Live), Thursday, November 18, 2010, broadcast time: 22:00–23:00.
3. Satoko Kuwahara, *Business is psychology—I will show you the backstage of a charismatic store where wisdom abounds*, KANKI PUBLISHING, 2005, pp. 188–191.

45

Provision of Organizational Information through Twitter

Atsushi Tsujimoto

Zappos' Wildly Popular Customer Service

The American Internet-based mail-order company *Zappos* (which handles shoes, apparel, accessories, clocks, etc.) achieved an annual sales figure exceeding 100 billion yen within a mere 10 years since its inception and is growing at a rate of more than 40% compared to the previous year. This is indicative of a super effect in terms of its business growth. In fact, its growth and management strategy have been hailed to be superior to those of the Internet mail-order behemoth Amazon.[1] In this section, I will elaborate on *Zappos*' *Twitter* effect from the perspective of Neo-Cybernetics.

"*Zappos* is the best!" "Their service was incredible", "Their delivery is supersonic", "I'm sure they've got a doctorate in customer service," "I'm writing as tears run down my cheeks…my gratitude can't be expressed in words, but thank you".[3]

These are the words of gratitude and praise sent by customers who had purchased the company's products. A large part of *Zappos*' brand value is found in the confidential relations the staff members build with their customers. So what is this service that customers rave about? The service that captures their heart? And how does it play out? Below I will speak about *Zappos*' *Twitter* effect, which has facilitated *Zappos*' communication with its customers and has also become a key tool for building confidential relations.

Zappos' Twitter Effect

"Build open and honest relationships with communication". This is *Zappos*' corporate slogan.[4] In this slogan we can see the company's enthusiasm for embracing the attitudes of "honesty" and "frankness" as they pertain to management conduct. To mention the words "honesty" and "frankness" in the official homepage — this mail-order firm's public face — a quality of bold readiness to stand behind those words is required. When considering how ardently customers praise the company to the extent of making comments like, "They get sincere and start offering advice on personal matters", you can see that the company is staying true to those words.[2]

The normative ideas and conducts of the company are also presented through the communication media that is *Twitter*. This media contributes towards building a relationship of mutual trust with customers by serving as a means of mediation between *Zappos* and its customers. It can be said to be a new corporate strategy/media strategy.

The CEO of *Zappos*, Tony Hsieh had his first encounter with *Twitter* at the 2007 SXSW Conference.[3] He noticed the effect of the medium to help build closer connections with friends and then realized that *Zappos* could use it also to build closer ties with its customers. For this reason, Hsieh adopted *Twitter* as a tool to communicate among staff members as well as a tool to communicate with customers. When customers or potential customers visit the *Zappos* website where the company carries out its mail order business, they are urged to join in a conversation with the company through *Twitter*, making them come into touch with the company's business attitudes and normative ideas.[2]

Some customers feel the conversations they have with the staff through *Twitter* are nothing more than "idle talk" and a waste of time. However this is actually "meaningful small talk" that offers the opportunity for both the seller and buyer "to know each other" and from the customer's perspective it is valuable time worth spending to understand the personal side of the *Zappos* salesperson in charge of selling to him or her.

Apparently first-time users who visit the site are able to read the flow of *Twitter* exchanges that take place between *Zappos* staff and customers. In effect, prior to making phone calls, ordering online or asking questions, first-time users are said to be able to gain an intimate sense of how reliable *Zappos*' operation is by reading through these *Twitter* messeages.[2]

It is here that I believe the neo-cybernetics observational space (complex communication structure) is inherent.

The Neo-Cybernetics Observational Space

The new trend in social system theory known as "neo-cybernetics" is a much talked about topic these days.[5] The basic line of thought in this field is a new observational information processing paradigm for information-processing-based cognition. This concept explains a novel observational space. While the univocal theme is "the subject observes his or her surrounding environment with his or her inherent emotions and mindset", the neo-cybernetics proposition is established when the subject's observation (or the act of observation itself) is observed by a different subject (observation of an observation). *Zappos*' sales and media strategy using *Twitter* is believed to be capable of explaining in particular the characteristic communication structure with the neo-cybernetics concept described as "observation of an observation".

During face-to-face selling, the way the seller faces the consumer is important. Since this company does not engage in face-to-face selling, a system that compensates for this situation is necessary. The seller who can convey the value of a product to consumers, or more importantly, the seller who can imagine the buyer's daily scenes to figure out what types of products could optimally satisfy the customers and then be able to communicate those views to them will enjoy regular customers.

Furthermore, the more these views are communicated with sincere affection, the stronger the relationship of mutual trust and the bond with the customer become. Generally, to build a relationship of mutual trust to a substantial level and to cultivate a strong bond, a considerable amount of time and effort is needed. In effect, *Zappos* is forming these very relationships of mutual trust and stronger bonds, which are conventionally cultivated through face-to-face selling, through friendly communication made possible *via* the above-mentioned medium, *Twitter*. In particular, with this form of communication, new visitors carry out value assessments regarding *Zappos* as a company.

This is a continuous act of "observing" *Zappos*. Due to the system configuration of *Twitter*, the communication process that takes place between staff members and customers can be "observed" as if from a bird's eye view. What should be noted here is that the neo-cybernetics observational space (complex communication structure) has been established here; that is, somebody somewhere who isn't communicating with a *Zappos* staff is carrying out the above-mentioned act of "observing an act of observation".

In other words, if this somebody somewhere gained a favorable impression of *Zappos* as a result of watching this communication process and consequently felt motivated to buy from the company, then in the final analysis, this neo-cybernetics act would have played a role in building a confidential relation/stronger bond with a new, invisible visitor.

To put it differently, the characteristic mass communication effect (the effect that causes secondary consumption behavior to arise) is inherent in *Zappos*' sales and media strategy using *Twitter*.

Note

*Another slogan that reflects the company's corporate philosophy is "Offer a moving experience through service". Additionally, the attitudes of "welcoming change" and being open to "learning and growth" are basic concepts that constitute the Theory of Organizational Learning, which has been attracting attention in recent years, while also being the characteristics of a sustainable management organization.

Bibliography

1. Shinobu Ishizuka, The Zappos Miracle (revised edition) — The most powerful management strategy in history that Amazon bowed down to, KOSAIDO PUBLISHING, 2010.
2. Umihiko Kotsukawa, Hiroaki Maeda (trans.), Business Twitter — The conversational 140 character media that changed companies around the world, Nikkei BP, 2010, pp. 140–143. (Shel Israel, Twitterville — How Business Can Thrive in the New Global Neighborhoods-Portfolio, 2009)
3. Alumni Association of Graduate School of Informatics, Kyoto University, *The Mania and Legend of Zappos*, Choukouryu Site (Johogaku.net). Available at http://www.johogaku.net/sn2011/program/sn2011s3 (20110611)
4. Zappos website. Available at http://www.zappos.com/ (20110611)
5. Clarke, B., Hansen, M.B.N.: Neocybernetic Emergence: Retuning the Posthuman, Cybernetics and Human Knowing, Vol. 16 (1–2), 2009, pp. 83–99.

Section 4:   Other Cases

46

Acceleration of the Commercialization of the Service Business in the Manufacturing Industry

Toru Fujii

Globally Unprecedented Entry into the Railway Business in Europe

In relation to the matter of updating high speed trains that interconnect major cities such as London and Manchester, the British Department for Transport announced on March 1, 2011 that it will enter into final negotiations with a consortium led by *Hitachi*, Ltd. (Chiyoda-ku, Tokyo, President Hiroaki Nakanishi) toward entering into a contractual agreement. Consequently, while the order was virtually received by *Hitachi*, the size of the business, which was at first said to be one trillion yen, was ultimately reduced by 40% to become approximately 600 billion yen (a total of 4,500 million pounds).

This scenario is a package contract that goes beyond the usual hard sell of selling only trainsets so as to include the selling of a maintenance service business for the duration of 30 years. In effect, the scenario is a super large-scale order amounting to approximately 600 billion yen. It became, for Japan, a key strategic arrangement for developing the export of package service infrastructures in the global market.

Brief Description of the Railway Business

Initially, *Hitachi* had obtained preferential negotiating rights from the British Department for Transport (November 2009) for a project that entailed whole surface replacements of diesel-powered high-speed trains, which had a vehicular career of more than 30 years. The order was large-scale, requiring the delivery of 1,100 to 1,400 trainsets within a five-year window from 2013 through 2017.

The business scheme guaranteed the availability of the trainsets for approximately 30 years and was positioned as a huge project with a budgetary scale that amounted to approximately one trillion yen for providing services that included the delivery of trainsets (approximately 100 billion yen for 1,100 to 1,400 trainsets), the provision of associated equipment, the implementation of overall engineering operations that included the construction of trunk lines, and furthermore, the provision of maintenance services.

However with the birth of the coalition government formed by the Conservative Party and the Liberal Party in May 2010, the reconstruction of public finances was pushed forward along with the mandate to re-examine various businesses; the rapid-transit railway business also fell under this classification. As a result, the assurance for the initially planned referential negotiating rights became uncertain and the bid was renewed. However, *Hitachi*'s efforts made during the past year had paid off and the company, which came to ultimately have preferential negotiating rights after all, won the bid and accepted the order at the price of approximately 600 billion yen (for 500 trainsets), which was 40% less than initially slated.

The factors that led to the receipt of this super large-lot order are as follows.

Factors that Led to the Receipt of the Large-Lot Order

In 2009, *Hitachi* had already received an order to provide 174 units of the Class 395 high-speed trainsets that allowed a maximum speed of 225 km/h, as shown in Fig. 46.1. Specifically, these trains were for servicing the Channel Tunnel Rail Link, a rail tunnel beneath the English Channel at the Strait of Dover linking the UK with France.

To develop these Class 395 trainsets, it was necessary for *Hitachi* to comply with international standards for European railroads and provide support for collision safety. As for international standards, there are two kinds: the

Fig. 46.1 Hitachi's high-speed train in the UK (Class 395)

Source: Yasushi Yokosuka and others, *Toward an optimal railway system that responds to a variety of needs with less burden on the environment.*

UK's "Railway Group Standards (RGS)/Rail Safety and Standards Board (RSSB) and Europe's Technical Specification for Interoperability (TSI)".

Each of these two standards are for their respective trainsets, and whereas the RGS is a set of standards conceived to cover all railway systems traditionally established in the UK, the TSI is premised upon mutual entry between various countries to open up the railway system to intra-regional nations. Furthermore, with regard to collision safety, in the case of Europe, it has been strongly sought after for ensuring the security of crewmen and passengers in the event of a collision between railway vehicles.

For this reason, research and development related to collision safety (including loading conditions, intensity of materials, aerodynamics, external noise, and fireproofing) has been traditionally carried out. In fact, in the UK, standards related to collision safety were established more than 10 years ago. For the project, the company had set out to work with a design based on the strictest standards, including the RGS for the UK, the TSI for Europe, and even the Japanese Industrial Standards (JIS).

With regard to *Hitachi* receiving the order for the Class 395 in 2009, the company was largely praised for following such a steady program of research and development, the achievements attained through the effort, and making delivery ahead of the requested deadline. Furthermore, with regard to the large-lot order received this time, the company was also highly praised for a point that is not surprising in Japan; due to the heavy snowfall in the UK during the previous winter, timetables of other trains became disrupted to a large extent, except for the Class 395 trains, which were running almost without any delays.

Furthermore, *Hitachi* also indicated that it will increase employment within the UK through building a new train factory in addition to proposing trainsets with a construction cost lower than the cost for constructing the initial trainset type. In addition to this, what worked favorably for *Hitachi* was the fact that the company had carried out top-level sales to a British governmental institution under the aegis of Japan's top governmental bank, based on the Japan Bank for International Cooperation's financing program proposal. The reason why *Hitachi* was able to obtain favorable treatment for this huge project over the world's major makers, *Siemens*, *Alstom*, and *Bombardier*, was because *Hitachi*, as a supplier, had proposed to take charge of all the facets of the business, offering, in effect, a "package type one-stop engineering service" that included the provision of overhead wiring work, maintenance, and operational control, in addition to the provision of hardware apparatus such as the electrical energy transformation system and signaling system for the main frames of rapid-transit trains, which have at their core the complex, globally peerless control technology applied in the operations of Japan's *Shinkansen* and *Yamanote* Line.

Of these, with regard to the maintenance service, as it is common for carriage makers to undertake the task in the UK, *Hitachi* established test trains and a train storage facility, for the purpose of performing maintenance work, at a site with a total area of 110,000 square meters that is contiguous to the Ashford International railway station. In the UK, there is no legally designated period for carrying out maintenance work, so such a period was to be proposed by the company carrying out the maintenance for approval by an inspection party. To realize profit growth, therefore, the key point became maintaining safe operations while raising maintenance efficiency.

By advising the *Ashford* railroad car base on how Japan's railway operations are maintained and introducing the period of maintenance work adopted by Japan's *Shinkansen*, *Hitachi* is ensuring safety through establishing a maintenance system with a shorter cycle than the one generally adopted in the UK. Specifically, the company is establishing:

1. Layouts specialized for the Class 395 trainset;
2. Inspections for optimizing work traffic diagrams;

3. Rail track placements suited for smooth interchanging;
4. Promotion of manpower savings;
5. Data management practices making full use of information technology.

For this reason, compared to the past, the company was able to realize decreases in defective vehicles and promote the efficiency of the maintenance system, making it now possible for them to anticipate fast profits from its maintenance service business.

Hitachi has already built a production plant for railroad cars in the northern province of Durham for the purpose of cultivating new markets in Europe.

Strategy for Exporting Infrastructure in the Future

While general electrical equipment manufacturers of the 20th century earned their profits largely from carrying out integrated engineering businesses on the basis of their strength in system coordination, the new revenue stream they are counting on from now on is "the export of infrastructures in the form of service packages".

In the case of *Hitachi*, its control technologies and its knowhow related to maintenance services for the company's client railroad company were achieved through the fusion of the two company's strengths in operational management (synergistic capabilities), and, in the end, is attributable to *Hitachi*'s valuable experience gained from co-creating with clients and suppliers over the long history of Japan's railway development to date. Furthermore, for the future, accumulating the know-how on the operational management and maintenance services carried out in the UK over the past 30 years will prove to be a competitive edge in the global market.

Bibliography

1. J-CAST News, broadcasted on March 12 (Sat) from 18:12 (20110505).
2. Toru Fujii, *New co-creative strategies through building synergistic capabilities,* Japan Academic Society of Hospitality Management, No. 18, 2011, pp. 77–87.
3. Yasushi Yokosuka, Takenori Wajima, Sumiyuki Okazaki, Tatsuo Horiba, Nobuyasa Kanekawa, Atsushi Suzuki, *Toward an optimal railway system that responds to a variety of needs with less burden on the environment,* Hitachi Hyoron, Hitachi, Vol. 89, No. 11, November 2007, pp. 13–19. Available at http://digital.hitachihyoron.com/pdf/2007/11/2007_11_01.pdf (20110505)

47

Improvement in Convenience and Economic Effects Achieved Through ID Sharing

(Expansion of Economic Effects)

Hiromichi Yasuoka

Explosive Growth of Linked Sites Through OpenID Support

Yahoo! JAPAN (Yahoo Japan Corporation, Minato-ku, Tokyo, President Manabu Miyasaka) made its *Yahoo*! ID system compliant with OpenID.* Consequently, it became possible for people with *Yahoo*! IDs of their own to have those IDs automatically authenticated to login at over 50,000 OpenID compliant sites. In other words, Single Sign-On (SSO) was made possible. Indeed, this is a drastic improvement in the level of convenience.

In this way, by just supporting OpenID, IDs** could be shared with other sites. As shown in Fig. 47.1, ID sharing not only enables SSO, but also makes it possible to share registered information. In addition, it makes it possible to share (or receive approval for carrying out) functions such as payments, conversion of points, and stock transactions. In the case of payments, there already are precedents set by *Yahoo*! and mobile carriers (au), while in the case of point conversions, there already are precedents set by companies such as *JCB*. Incidentally, in these cases, support for authorization is usually made with the OAuth* standard.

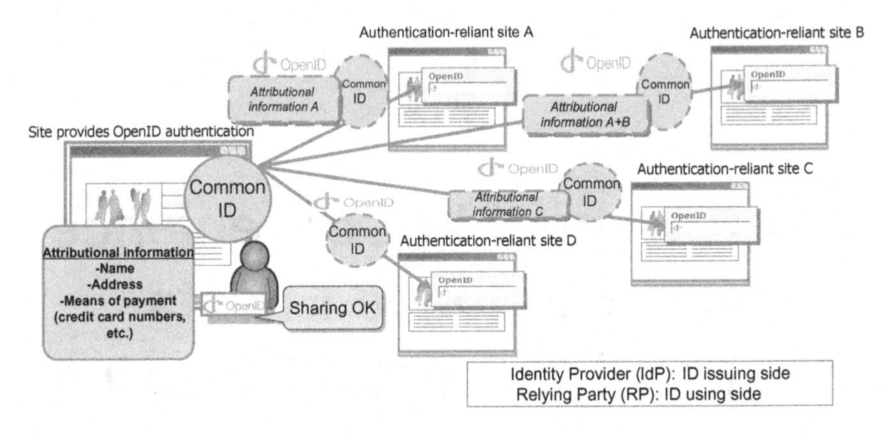

Fig. 47.1 OpenID's ID sharing

Source: Author.

In this way, the business is developing into one that offers value for enabling the sharing of various information and functions between firms issuing IDs and the firms making use of them.

In Japan, from February 14, 2007, OpenID.ne.jp became OpenID compliant, and in May 7 of the same year *Livedoor* followed suit. Since *Yahoo*! Japan also became compatible with this standard on January 30, 2008, the IDs it issued also became shareable among domestic websites for the purpose of logging in. In addition, *Google*, *Hatena*, *mixi*, *BIGLOBE*, *Excite*, *Rakuten*, *NTT* (*docomo*, *goo*, *OCN*), *JAL*, *Twitter*, and *Facebook* also became OpenID compliant.

At JAL, support for OpenID started on May 28, 2008. Until then, users logging into *JAL*'s website were required to log in with their *JAL* Mileage Bank membership numbers, and in the case of *Myu* (MIKI TOURIST), an overseas hotel reservations website that is affiliated with *JAL*, users were required to use another set of ID numbers separate from their *JMB* membership numbers. By having JMB membership numbers become OpenID compliant, Myu began to accept those numbers as their login IDs. Furthermore, if a user's credit card information were registered at *JAL*'s website, that same information could be used to complete payments for hotel reservations by linking the credit card information from *JAL*'s website. Furthermore, the company has begun to have similar linkups with *JAL* hotels from October 1 of the same year and has a swift increase in the number of users registering to this service.

We are also seeing a rise in the number of websites offering OpenID support. Users accessing *Ito-Yokado*'s website can log in using *Yahoo*! ID and users accessing *e.design sonpo*'s website (TOKIO MARINE GROUP) can log in using IDs issued by *NTT*. Furthermore, users accessing *Recruit's SUUMO* can log in using their *Twitter* or *Facebook* IDs. As can be seen, the OpenID standard is being used by not only Internet startups, but by many conventional companies (belonging to many different industries).

In addition, we are beginning to see the appearances of companies overseas not only in the online world, but also in the brick and mortar world that are starting to make their card numbers OpenID and/or OAuth compliant to facilitate authentication and information sharing.

In this way, companies can allow not only users to log in with the IDs they issue, but also with the information associated with the IDs issued by another company. In essence, if the user consents, it becomes possible to share such information anywhere. In other words, it becomes possible to automate re-entering and updating ID data by enabling the sharing of information linked to user IDs dispersed ubiquitously (for example, one-click notification of an address change when relocating), and improve the accuracy of information provision by showing information that users seek (for example, collecting recommendations that further meet one's needs).

Incidentally, the cooperative sharing of such information (personal data) began to attract attention across the world, and at the *Davos Forum* (World Economic Forum), this cooperative use of information was named the currency of the digital world.

Effective ID Sharing in Japan, a Nation Inundated with IDs

Compared to other advanced countries, the proliferation of IDs in Japan is extremely great, making it appropriate to call the nation an "ID-inundated country". According to a 2008 survey carried out by *NRI*, men in their 30s, the bracket that owned the most number of Internet service IDs, had 8.5 different types of IDs (different business categories), and women in their thirties, the bracket that owned the most real IDs (actual cards), had 9.2 different types of IDs.[1,2] The actual number they possessed was more, easily surpassing 20.

The reason why companies that issue IDs attempt to increase the number of IDs (the number of members) in various ways is because increasing ID numbers actually leads to improving the value of the firm. Once the number of IDs increases, the quality of the information linked to the ID goes up (of course, this will depend on the content of the information), elevating the accuracy of customer information analysis. By using this analysis to make management decisions, it is thought that the probability of success will rise.

Consequently, the movement to tie up with other companies to have them contribute IDs (members) was created as an alternative to solely relying on IDs issued in-house. In addition, the business of cooperatively sharing IDs (authentication sharing, introducing customers, and even functions such as facilitating payments) gained ground by carrying out targeted advertisements and recommendations through leveraging the information linked to these IDs.

Furthermore, a company's services can now be used as if they were the same services of external services (other companies). For example, there already exist hookups that make it possible to carry out stock trading at a securities company site (e.g. *E*TRADE* in the US) from external sites.

The Effect of Tie-Ups to Share Private IDs with Administrative Services

Based on the trends of the times mentioned above, we should permit the use of IDs issued by private enterprises (enable cooperative ID sharing) for the purpose of accessing governmental services, rather than allowing such access with only those IDs issued by government offices. In so doing, it will become possible for residents to easily use administrative services, and it will also become possible to align the foundations of administrative services more quickly and inexpensively.

To easily understand the way these private IDs are used, an analogy with one's seal will help. For example, what corresponds to one's registered seal is the governmental ID issued by local governments (common numbers for resident registry IDs, social security to be introduced and tax). On the other hand, there also exist private bank IDs (in the case of seals, this corresponds to the bank's seal) and IDs for various sites and points programs (in the case of seals, this corresponds to the private seal). Therefore, these IDs should be appropriately used and shared in accordance with the level of personal ID authentication necessary for the purpose of accessing various administrative services (receipt of child allowance, acquiring a resident card, application for library borrowing privileges). In doing so, residents will be able to effectively make use of administrative services. In the US and Netherlands, the governments there have already introduced a framework of trust, which is spreading, that categorizes administrative services into four levels of security, allowing for even people with private IDs to carry out personal ID confirmations.

Furthermore, "improvements in convenience levels/realizing efficiency" and "business creation" could be anticipated by applying this type of framework of trust that differentiates security levels to the domain of private enterprises. The impact this would bring about is estimated to reach the economic effect of up to approximately 10,500 billion yen,[3] which will be a super effect indeed.

Notes

*One of the key standards that enables multiple services to share user profiles in common. When a user registers with an OpenID compliant site once, he or she can register at other OpenID compliant sites without registering again, using the same ID to log in. At the same time, sites can share registered information (full name, address, credit card numbers, etc.) with the user's approval. There already are over 50,000 sites in the world that support OpenID with more than 1,500 million OpenID compliant IDs issued and in circulation. OAuth is a similar standard that is suited more for the sharing of functions.

**ID refers to "personal identification" and "personal identity," which also includes the information linked to an ID number, and in this part of the book ID refers to the latter definition.

Bibliography

1. Nomura Research Institute ID Business Project Team, *ID business in 2015 — Facilitating a wide range of functions, ranging from vending machine purchases to public authentication with just one card,* Toyo Keizai Shinposha, 2009.
2. Nomura Research Institute, *Industry Reorganization — Now & Future No. 22, "Collecting and Sharing IDs,"* Nikkei Publishing Inc., 2010.
3. Nomura Research Institute, *The Effect of Introducing the ID Ecosystem — Leveraging the strength of private enterprises for the national ID system,* The 148th Media Forum, February 21, 2011. Available at http://www.nri.co.jp/publicity/mediaforum/2011/pdf/forum148.pdf (20110627)

48

The *Blue Card's* Customer Attraction Effect in Nagano Prefecture

(Customer Attraction Growth)

Hiromichi Yasuoka

The *BLUE CARD*'s Principal Partnerships and Requirements for Participation

There are points programs that are superbly capable of pulling in customers within local communities, unlike the national point-card schemes. One of them is the *BLUE CARD* (BLCA), a point-card scheme for local communities run by *BLUE ALLIANCE* Co., Ltd. (Minato-ku, Tokyo, President Kazuya Kawano). In Nagano Prefecture in particular, the points program run by an affiliate, *Shinshu Communications*, Co. Ltd. (Nakagosyo, Nagano-shi, President Minoru Hirano) has a high customer attraction effect.[1]

This local association of point-card systems of Nagano was established in June 1986, and as of March 2011, it already had a membership of 500 affiliate stores. The affiliated stores include not only stores offering consumer goods and transportation, such as home centers, supermarkets, bookstores, opticians, shoe stores, apparel, department stores, a taxi service etc., but other services — basically there is one company per industry type participating.

In addition, the number of *BLUE CARD* cards issued has exceeded 560,000 with the number of valid members exceeding 440,000. In terms of valid members, this is around 45% of Nagano's household diffusion rate. The handling amount (distribution total) has reached the scale of 60 billion yen.

Other than Nagano, the program is also running in *Iyo* (Ehime), *Yamaguchi*, *Aichi*, *Yamagata*, *Niigata*, *Oita*, *Kitakyushu* (Fukuoka). The company plans to carry out this program by specializing in local communities from now on as well, distinguishing itself from programs like *T Point* and *Ponta* that have tie-ups with mainly nationwide chain stores.

Although the basic monthly management charge for affiliate stores participating in the point-card system applies to each of them on a pay-per-use basis, there is a fixed upper limit. In addition, a fee worth 1% of a point granted (in the case of double, then that sum will be charged) and a handling fee is charged occasionally and the points used are realized on the tenth of the second month, offsetting the points granted.

As another option, the input data of sales slips and purchase histories (subscriber ID + the JAN code of POS data) could be used for promoting affiliated stores. For example, it is possible for a wedding party facility to inform about wedding fairs to consumers who have purchased magazines on marriage information at bookstores.

The *BLUE CARD* Customer-Gathering Effect

As shown in Fig. 48.1, *BLUE CARD* members are basically granted a point worth 1% of the amount of a purchase made from the above-mentioned affiliated stores and are able to use the points at those stores at the rate of 1 yen per point. However, these points are not available to them right away. Instead, they are mailed a 500-yen coupon for goods (worth 500 points) every six months. This is the form in which they are able to use the points and if they do not cash them in within two years after receipt, the points expire.

Despite such terms, the utilization rate in northern Shinshu is enormously high at 94 to 97%. Normal points programs often allow usage of the points from one yen, but even in such cases, this utilization rate of 90% is still considerably high.

The fact that the utilization rate of the points program is high means that the program is highly effective in promoting customer traffic to participating affiliate stores. Since realizing a 1% raise requires considerable

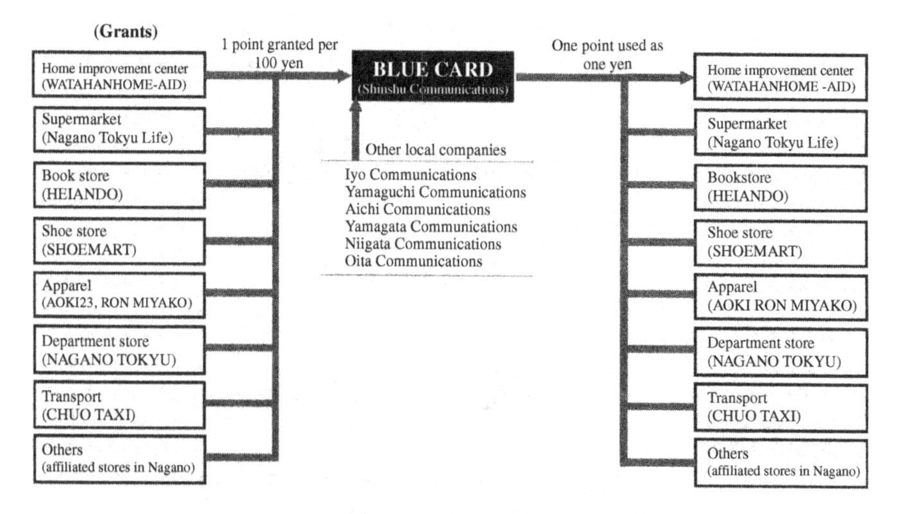

Fig. 48.1 Nagano's association of the *Blue Alliance* regional points program

Note: Most of these are returns to the original grants.

Source: BLUE ALLIANCE website and hearings conducted by the Chapter's author.

measures, and since the utilization rate is 14% to 17% higher than usual, which is 80%, the effect can be said to be a 14 to 17 fold super effect.

BLUE CARD's Marketing Action Plan and its Future

BLUE CARD is structured in such a way so as to make it conducive to carrying out marketing programs that are tailored to the local stores. For example, an affiliated supermarket can carry out a measure such as doubling the points as a feature of a special time sale (sales differentiated by time zones and products). Such a measure is extremely effective *vis-à-vis* the main customers: housewives in their 30s through 50s. In addition, a customer line forms by just placing a "*BLUE CARD* flag" over the traffic line.

Forming an association that caters to the special needs of a local economic bloc in this manner also proves to be effective in activating the regional economy. By cooperating, it becomes possible to hold down indirect expenses as much as possible and put into effect measures that can heighten buying inclinations revolving around real transactions that all levels of community residents can take advantage of.

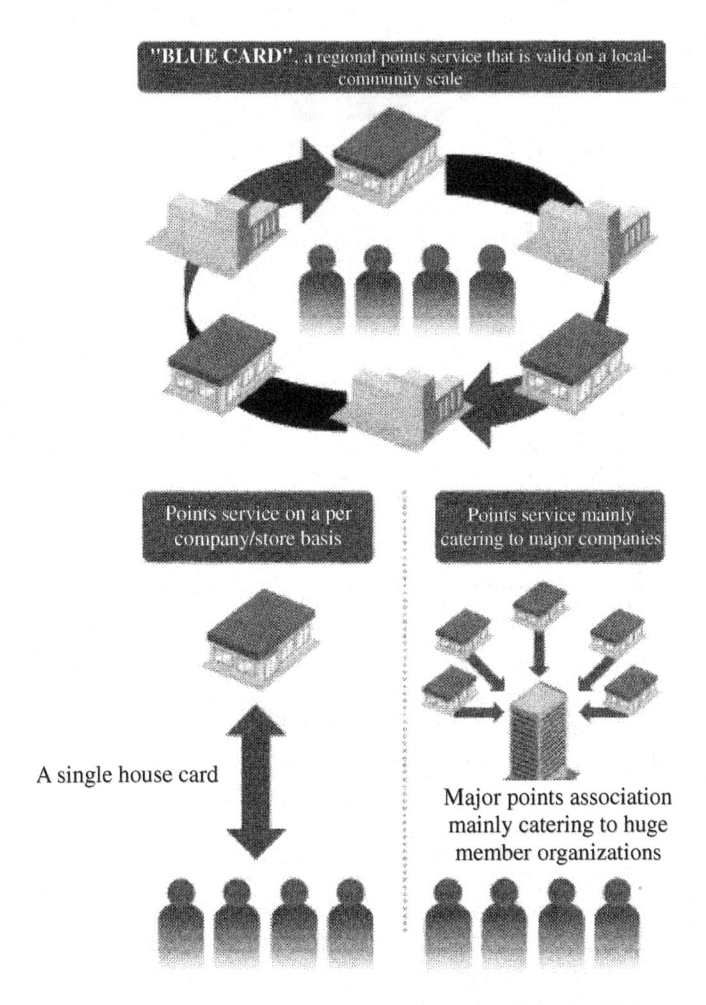

Fig. 48.2 Prefectural points association (*BLUE ALLIANCE*)

Source: *BLUE ALLIANCE* website.

Therefore, as shown in Fig. 48.2, a community-based points program, as opposed to the types of programs offered by national chains, is offered instead of a program that provides on a company by company basis.[2] Furthermore, *BLUE ALLIANCE* itself is attempting to acquire know-how of cards that offer common points in the UK such as Tesco and Nectar to propose lifestyle ideas for members gained from this know–how.

The key aspect about this regional points association is that it tends to propagate cooperation within the local community by carrying out local promotions in opposition to national chains whenever such chains open stores in their local communities.

For example, while there exist more than 600 regional currencies issued by local communities, for the most part, they are considerably restricted to limited areas, and are devoid of any flexibility even within prefectures, ending in failure as a result.

In other words, unlike general-purpose electronic monies like *Suica* and *Edy*, newly issuing currencies that are unfamiliar to begin with and that are unusable in other communities even if they happen to be within the same prefecture is a recipe for failure, driving home the point that some flexibility is necessary. Resolving this problem was none other than the *BLUE CARD* local points association.

From now on, if media that represent local communities are also taken in by affiliated stores (advertising agencies, newspaper companies, Internet Service Providers (ISPs) of local communities) and the concept of local production for local consumption is promoted effectively, it will further catch on within a local community as a whole. Furthermore, if we start to include elements of regional sightseeing, such as hot spring resorts and world heritage sites, there is a likelihood that the initiative will develop into a local community-wide cooperation.

Bibliography

1. Shinshu Communications website. Available at http://www.bluecard.ne.jp/ (20110831)
2. BLUE ALLIANCE, *What is BLUE ALLIANCE?*, Business information. Available at http://www.bluealliance.co.jp/service/bluealliance.html (20110831)

49

Increasing the Number of Visitors Through Twitter

Enlarging the Number of Followers

Junpei Nakagawa

Achieving Regional Revival with *Twitter*

"Japan's retail areas" used to refer to shopping centers found in station plazas that comprised of many individual stores. However of late, these types of areas have been waning due to the graying of society and the proliferation of automobiles. Nonetheless, we are now seeing instances of visitors increasing by putting *Twitter* to practical use for local revivification. Here, I would like to examine the significance of shopping streets making use of *Twitter* as an organizer for individual stores by referring to the pioneering example of the *Koenjilook* Shopping Street in Suginami-ku, Tokyo. This shopping street has already helped increase the number of followers to more than 6,000 and is actively pursuing a *Twitter* linkup by promoting the use of *Twitter* to each individual store.

The *Koenjilook* Shopping Street extends between the south exit of Koenji Station accessed by the JR Chuo Line to the Shin-Koenji Station accessed by the Tokyo Metro Marunouchi Line. Unlike retail areas such as Shinjuku and Kichijoji, which teem with large stores such as department stores and general merchandise stores, Koenji Station is a community-based retail area comprised of individual stores and small-scale chain stores. In particular, the area is home to several shopping streets, including the Koenji Junjou Shopping Street, which has received extensive coverage by the mass media, including TV. For the purpose of local revivification, Koenji has been actively engaging in measures such as annually hosting the event of the Awa Folk Dance. Koenji has been successful in this effort despite the fact that the Awa Folk Dance is a festival native to distant Tokushima, not to mention that it is also alien to Tokyo culture.

However, if we exclude the senior bracket that was instrumental in building long-term social relations within the local community, we begin to see a shift of customer traffic from individual stores to large-scale chain stores, and this trend isn't restricted to just local cities but can be seen within Tokyo as well. Consequently, stemming this tide had become a challenge for a long time.

While the *Koenjilook* Shopping Street regularly invited comic story tellers and held the "Look Vaudeville" event in an effort to achieve regional vitalization, when the shopping street began to report on the actual scenes of this event in real time *via Twitter*, its initiative suddenly began to attract the attention of young people.

The Interactivity of a *Twitter* Account

What should be highlighted in this example is the fact that the shopping street's account, the accounts of each store, and the accounts of individual users are organically related to each other, showing how effectively *Twitter* is being utilized. At the *Koenji* Shopping Street, there were many stores that were suited for using *Twitter*, such as boutique restaurants or select stores handling one-of-a kind items (e.g., secondhand stores and general stores). In addition, compared to other shopping streets, many of the proprietors of the stores belonged to the young generation so many of them had no resistance to the notion of using social media. For this reason, various establishments in the shopping street went on to actively acquire *Twitter* accounts of their own to actively transmit information.

With stores run by individual proprietors, the effect of sending direct mail to an unspecified number of local residents is often minimal, failing to justify the huge costs entailed in the process. In addition, since only one or a few people necessarily run these stores, establishing and running a blog would prove to be too time consuming. Consequently, the use of *Twitter* helped to resolve these two demerits, but the challenge remained of how to go about overcoming the low recognition levels of the stores themselves to increase the number of their followers.

Therefore, calls began to be made for the Shopping Street to set up an official account to play the central role of an organizer for individual stores. Linked to the *Look* Shopping Street's account (*@koenjilook*), the shopping street's website was also established, owing to the fact that Mr. Nakazawa, a web designer, was residing in the shopping district.[1]

The *Look* Shopping Street's account carries out an organizer-like role, retweeting the tweets from the accounts of individual stores and pointing to relevant stores in response to inquiries tweeted from followers. In addition, if we take a look at the *Look* Shopping Street's webpage, it is possible to observe tweets on social media feeds being streamed one after another from either of the shopping street's accounts.[2]

Shoppers who frequently visit the shopping street have already become earnest followers and whenever they find a stock of new products or unusual items, they frequently tweet about them, indicating which stores they are available at, and the *Koenjilook* Shopping Street also instantly follows up on them. As a result, as shown in Fig. 49.1, there is already a structure in place that allows anyone following *Look* Shopping Street's account to catch timely, seasonal information related to the shopping street anytime.

And even in this example, there can be seen many casual posts from senders that have a sense of immediacy about them. For instance, at the

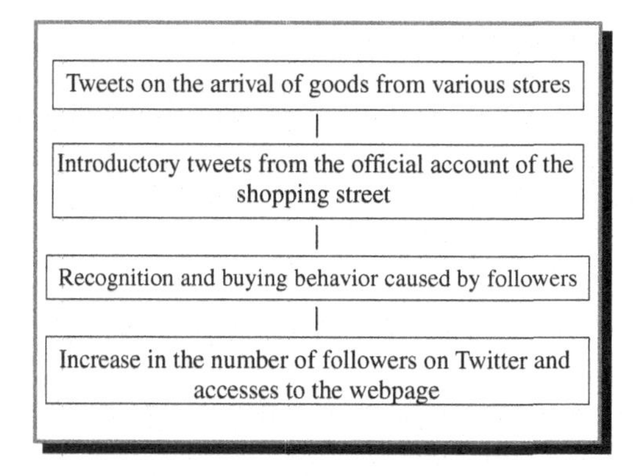

Fig. 49.1 The *Twitter* effect produced by the Koenjilook shopping street
Source: Author.

time of the outbreak of an active aftershock of the Great East Japan Earthquake, wherever people immediately tweeted descriptive prose of how violently they felt the earth shaking, since people in the same regions were simultaneously experiencing the tremor, these people were automatically driven to look at such tweets to understand just what the magnitude of the tremor was like to the senders of those tweets. In essence, tweets have a mechanism that causes such compulsion.

The *Koenjilook* Shopping Street has been holding personal computer classes to help introduce *Twitter* to veteran shopkeepers who have had shops in the shopping district for a long time, and have also been urging the shopping districts in Suginami-ku and other regions to introduce the medium. Consequently, these places took the advice and have started to show highly effective results.

The Spread of *Twitter* into Other Shopping Districts

The *Nishiki* Market in Kyoto, as part of a university-industry research collaboration with Digital Hollywood, a designer school in Osaka, has updated their webpage under the student OJT program and established an easily likeable mascot while also initiating *Twitter*-account tie-ups with various stores.[3] The organizers have set up the *Nishiki Market's* website so as to allow visitors to read tweets from various stores just like the website of the *Koenjilook* Shopping Street, and it is said that the visitor number has risen ten-fold. Compared to the *Koenjilook* Shopping Street itself, the number of followers of the *Nishiki Market* account's is still small, but the organizers carried out a live broadcast relay *via Ustream* of the "Pan Festival" held in the winter, and more recently, held an event titled "Let's drink *sake* from the northeast in the Nishiki Market" as part of an effort to support the disaster-stricken areas. Offering tidbits of Kyoto cuisine such as *dashimaki* (a Japanese style omelet), deep-fried foods, and dried bean curds, the event, which promoted the consumption at the shopping district of local brews from the northeast, was relayed from the spot *via Twitter*, enlivening the venue.[4]

With the tweets of thanks from shoppers from Tohoku (the northeast) also posted on the *Twitter* pages of stores, the event turned out to be a significant one that put together the regional vitalization of Kyoto and the area rehabilitation of the Tohoku region. With the theme of "supporting Kyoto's kitchen," *Nishiki* Market is a local resident-centric shopping street, but by staging regular events, it has attracted many visitors from distant places, just as we saw in *Koenjilook's* example.

While shopping streets located within station plazas are on the decline, the notification technique that leverages the localness of *Twitter* will prove to be a trigger for reviving local communities.

Bibliography

1. ITmedia News, *Why are shopping districts tweeting? — We ask Koenjilook, the front runner that pioneered the use of Twitter for shopping districts,* ITmedia, February 19, 2010. http://www.itmedia.co.jp/news/articles/1002/19/news066.html
2. Koenjilook Shopping Street website. Available at http://koenjilook.com (20110531)
3. Nishiki Market website. Available at http://www.kyoto-nishiki.or.jp (20110531)
4. Karasuma Keizai Shimbun, *Charity event at Nishiki Market,* May 16, 2011. Available at http://karasuma.keizai.biz/headline/1389 (20110531)

50

Energy Self-Sufficiency with Algae that Produces Oil

Discovery of a Raw Material

Kazuo Matsude

The Difficulties of Japan's Energy Strategy

The aim of this chapter is to provide an opportunity for contemplating the impact made to Japan's energy strategy by introducing a case example of an innovation that is having a super effect in the field of energy development.

In its 2006 report titled "The New National Energy Strategy", the Ministry of Economy, Trade and Industry presented a summary about reviewing the energy situation until 2030 and Japan's basic policy on the situation. From the outset, the report declares the goal of reducing oil dependence, which stands currently at approximately 50%, to a level below 40% by the year 2030, and to this end, the report proposes the engagement of the following four plans.

1. The energy-saving front-runner plan (Improve energy efficiency by a further 30% by 2030)
2. Next-generation transportation energy (Aim to reduce oil dependence, which is currently approximately 100%, to around 80% by 2030)
3. Energy innovation plan (Make the cost of photovoltaic power generation at the same level as the cost of thermal power, support measures

that promote local production for local consumption utilizing biomass energy, raise the regional energy self-sufficiency level, etc.)

4. National nuclear plan (Maintain a 30% to 40% share of generated energy even after 2030 and establish an early-stage nuclear fuel cycle to engage in the practical use of the fast breeder reactor at an early stage)

However, due to the radiation accident of the Fukushima I Nuclear Power Plant triggered by the 2011 Tohoku earthquake and tsunami disaster, renewed skepticism over the safety of nuclear power supply arose, and a drastic reassessment of Japan's nuclear energy administration became unavoidable.

With the prospects for plan number four mentioned above uncertain now, creating the perception that the final goal will be difficult to achieve, the importance of filling the gap with plan number 3 can be said to have heightened all the more.

Super Effect of Algal Biomass

On December 14, 2010, at The 1st Asia-Oceania Innovation Summit, co-sponsored by Tsukuba University, the Japan Science and Technology Agency, and the Cabinet Office, Tsukuba University's professor, Makoto M. Watanabe and others announced their discovery of *Aurantiochytrium*, a promising strain of algae that could harvest biofuel.

The production efficiency of this algal liquefied petroleum gas is said to be 10 times more than the production efficiency of *Botryococcus*, which was the strain with the best production efficiency until the discovery was made. As shown in Fig. 50.1, Professor Watanabe said, "In terms of oil production, *Aurantiochytrium* is 1/3 that of *Botryococcus*, but its good point is the fact that it has a speed of propagation that is 36 times faster. A simple calculation reveals that the production efficiency, compared to the past, has increased twelve-fold". This is indeed a super effect.

Through photosynthesis, *Botryococcus* fastens the carbon dioxide found in the atmosphere and makes hydrocarbon from nitrogen and phosphorous found underwater. On the other hand, *Aurantiochytrium* does not photosynthesize but since it has the ability to create hydrocarbon by dismantling the organic matter of sewage or sludge, this strain is classified as a biomass resource, the same category to which rapeseed and soybean

Maize	0.2t/ha	
Soybean	0.5t/ha	
Oil palm	6t/ha	
Botryococcus	>100t/ha	
Aurantiochytrium	>1,200t/ha	(In the case of converting into plants that photosynthesize)

Fig. 50.1 Comparisons of biofuel production efficiencies (in a year)
Source: Data extracted by the author from Tatsuya Yamaji's *Questions for Eco-technology researchers — Aurantiochytrium will make Japan an oil producing country.*

belong. (Note that biomass resource here is defined as "a material that is well organized as energy and has an origin in plants).

According to provisional estimates, the research group notes that the production cost to convert *Botryococcus* into liquefied petroleum gas cannot currently be more than 800 yen per liter whereas in the case of *Aurantiochytrium*, this same production cost could drop to around 50 yen per liter with large-scale cultivation. If this is achieved, the strain will come to have sufficient price competitiveness over current crude oil prices, and since this prospect makes the strain a viable commercial product, it has come into the limelight.

Moreover, the cultivation of algae has the significant merits of being unaffected by rises in food prices and being detached from ethical concerns for famine-stricken areas, which are issues that accompany the conversion of food crops such as rapeseed or soybean into liquefied petroleum gas.

The Business Model of Algae Biomass

For an oil production facility using *Aurantiochytrium*, incorporating tanks in the treatment facility for processing organic sewage that will become bait will allow processing of sewage and the production of oil at the same time, and this method is thought to be the most cost-efficient as well.

In other words, while it is common in the biologic treatment of sewage disposal to dismantle mineral matter such as water, carbon dioxide, phosphoric acid, and nitrogen by metabolizing organic matter of primarily treated sewage (arrived at by removing solid bodies from the organic sewage) with an activated sludge tank using microbes during the secondary treatment, if the metabolic process of *Aurantiochytrium* can be added into the process, we could kill two birds with one stone.

Currently, reductions in the level of industrial waste can also be anticipated, since the excess sludge generated annually has reached 400 million tons and there is no uncertainty regarding the raw material aspect for the time being. Furthermore, since urban areas mainly serve as locations of sewage disposal facilities, transport costs can also be held down.

The Future of Algae Biomass

With the discovery made this time, algae biomass is currently at a stage of being recognized for the first time as a major candidate for profitable commercial production, but there still remain some technical issues to overcome in order to achieve its practical use; applying the churning method or the oil extraction method to achieve the efficient production of algae remains a challenge. However, it is important to note that the profitability of algae biomass is now in sight, and for this reason, this raw material should be able to spur the diffusion of innovation.

In addition, if we heed the warnings of the difficult energy situation Japan has been facing lately, we can expect strong backing for this project from both the industry and government sides to be forthcoming.

Bibliography

1. Tatsuya Yamaji, *Questions for Eco-technology researchers — Aurantiochytrium will make Japan an oil producing country 2,* WIRED VISION, February 25, 2011. Available at http://archive.wiredvision.co.jp/blog/yamaji/201102/20110-2251302.html (20120929).
2. Makoto M. Watanabe, *Algae biomass: a new energy resource,* IGAKUHYORONSHA, 2010.
3. Masato Kagawa, Nihon Keizai Shimbun Electronic Edition, *Fundamental research into the viability of algae as an oil substitute advances into next-generation industry,* December 24, 2010.
4. Japan Science and Technology Agency, *Coverage of the 1st Asia-Oceania Innovation Summit,* JST Topics, Japan Science and Technology Agency, January 18, 2011. Available at http://www.jst.go.jp/report/2010/110118.html (20110410).
5. Ministry of Economy, Trade and Industry, Agency for Natural Resources and Energy, *New National Energy Strategy,* Ministry of Economy, Trade and Industry, 2006.

6. Tatsuya Yamaji, "Questions for Eco-technology researchers — Aurantiochytrium will make Japan an oil producing country 1," WIRED VISION, February 25, 2011. Available at http://archive.wiredvision.co.jp/blog/yamaji/201102/20110-2251301.html (20120929)
7. Tomoyuki Yamamoto, Asahi Shimbun Electronic Edition, *"A promising strain discovered in Japan — Algae, a means to make oil with a production capacity that is ten times more,"* December 15, 2010.
8. Makoto M. Watanabe, "Algae biomass: a new energy resource," IGAKU-HYORONSHA, 2010.
9. Society for the Study of Sewage Business Management, *Resourceful management and administration for businesses,* SANKAIDO, 2006.

51

Negative Effects of International Politics

Akira Ishikawa

With regard to Japanese defense power, which is also closely associated with international politics, I would like to point out just how much it differs from the defense capabilities of its neighboring countries.

As a conceptual framework to maintain international peace and order, the theory known as Balance of Power was made much of in the field of international politics, in particular after the 19th century. Mainly championed by the UK, it was inherited in the 20th century when the Cold War was taking place, and to this day, it remains influential.

The results of the research conducted by Organski, Kugler and others are well known. Upon examining the Franco-Prussian War, Russo-Japanese War, World War I, and World War II and analyzing the conjectured national powers of nations 20 years prior to the outbreak of each of those wars, the political scientists found that the incidence rate of a war occurring was more than 50% between nations whose military powers varied by more than 20%.

In that case, what would we see if we compare national powers between Japan and its neighboring countries, particularly in the area of defense capabilities?

As partly shown in Diagram No. 1 of the 2010 Defense White Paper (5 pages) appearing in Fig. 51.1, North Korea's armed forces, which was responsible for attacking Korea's Yeonpyeongdo Island, killing many, including private citizens and soldiers, is estimated to comprise of an army of one million soldiers (27 divisions), an air force of 620 planes,

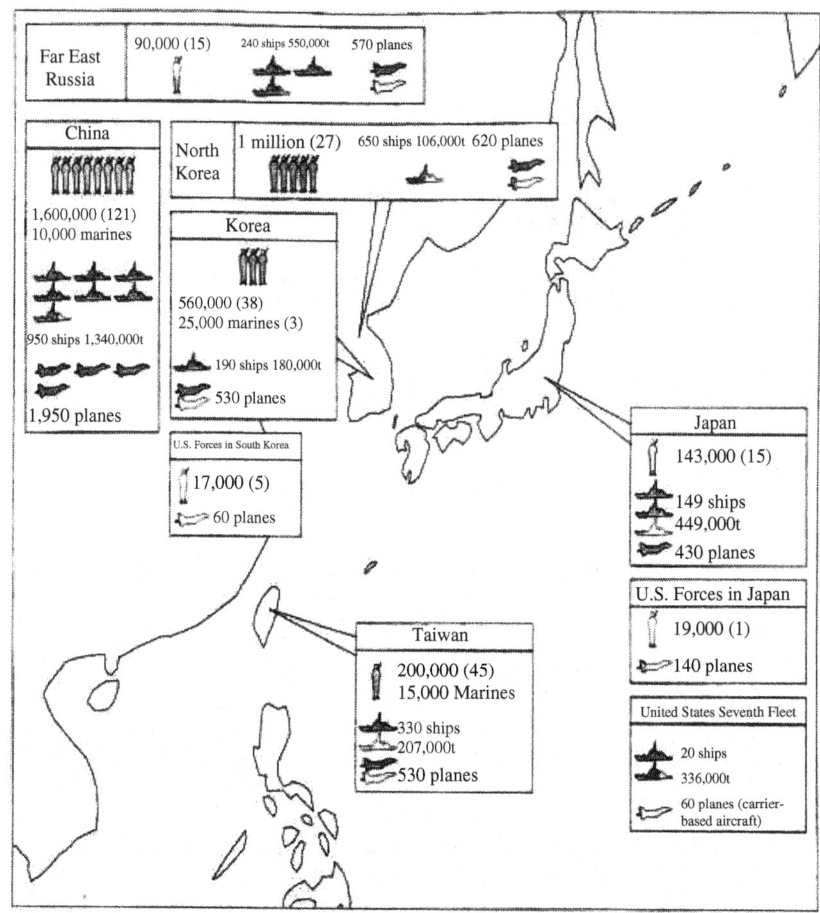

Notes
1. These datasets are extracted from the publicly disclosed materials of the US Department of Defense, including one titled Military Balance (2010) (For the end of 2009, the fi gures for Japan reflect its actual forces)
2. The land-based military force of the United States Armed Forces stationed in Japan and Korea refers to the total number of soldiers in the army and the Marine Corps.
3. With regard to combat aircraft, naval carrier-capable are included.
4. The figures in the brackets indicate the total number of basic military units, such as divisions and brigades. With North Korea, only divisions are reflected. With Taiwan, MPs are included.
5. With regard to the United States Seventh Fleet, the indicated military forces are those that have forward deployment in Japan and Guam.

Legend

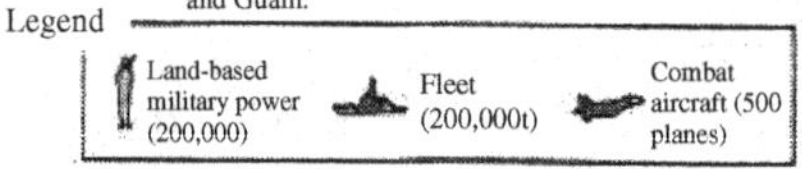

Fig. 51.1 The overall situation of the military powers in the Asia-Pacific region

Source: Ministry of Defense, *Intensification of international competition and the need for innovation,* 2010, Defense White Paper.

along with 63 submarines and more than 10 nuclear warheads with at least 10 nuclear facilities and missile bases in eight places. The military is also in possession of dozens of missiles named *Taepo Dong Type* 2 and *Type 1* (which are estimated to have flying distances of 6,700 km), the *Nodong*, the *BM25 Musudan* (flying distance of 3,000 km), and the *Scud-B&C* (flying distances of 300 km–650 km).

Of course, China's military capabilities are incomparably more than North Korea's, as it has a total of approximately 1,600,000 soldiers, 10,000 marines, a naval force made up of 950 ships, an air force made up of 1,950 planes, along with 46 Intercontinental ballistic missiles (ICBMs), 35 Intermediate-Range Ballistic Missiles (IRBMs), 725 Short-Range Ballistic Missiles (SRBMs) in addition to 12 submarine launched ballistic missiles. Clearly, China's armed forces are far superior to North Korea's forces.

Furthermore, with regard to Russia, even if we underestimate its capabilities, we can say that it possesses approximately 5,600 nuclear warheads, 489 ICBMs, 180 SLBMs, and even if we don't take into account the rest of its military power, the nation secures a military capacity that far outstrips that of China.

But that is not all. In both India and Pakistan, the number of nuclear warheads that Japan lacks is estimated to be at least 50. In other words, Japan's military pales in comparison to the total military power or nuclear capabilities of neighboring countries in East Asia whose sizes show a double-digit difference.

From the perspective of the Balance of Power theory, Japan has been totally dependent on the United States to counter the most important part of this extreme asymmetry, while walking the extremely dangerous tightrope of having to trust the good will of its neighboring countries.

Amid such a military milieu, Japan has no other options for survival but to fully trust the security provided by the American nuclear umbrella and develop super-weapons far superior to nuclear weapons, or increase the nation's information gathering capability to an extremely high level.

One potent and simple means to strengthen and maintain this capability is to utilize a sufficient number of information-gathering satellites. Currently, Japan's information-gathering satellites are in orbit at a height that is 400 km to 600 km higher than that of American satellites, but then it is still necessary for Japan to have its satellites orbit at the ultra-high altitude of approximately 20,000 km like the global positioning satellite cluster and communication satellite networks, or add self-defense capabilities to counter-attack or evade enemy attacks, or continually prepare complimentary satellites.

Furthermore, in the event of an emergency, Japan must be in a state of readiness by developing a technology that can destroy satellite clusters orbiting at ultra-high altitudes. As some commentators on military affairs have suggested, the need to establish an Aerospace Self-Defense Force is necessary to aim for using the realm of outer space.

The reason for this is that if these satellite clusters and satellite networks were to be destroyed, guiding precision smart bombs and missiles, and determining positions of aircraft and vessels will become difficult, making it impossible to successfully wage modern warfare, a system that is linked to computer networks.

Note

Chapter 3 of Part I, Chapter 16 and Chapter 23 of Part II, Chapter 36, Chapter 42, and Chapter 51 of Part III are all based on essays in the "Seminar" column of THE NIKKAN KOGYO SHIMBUN. Their respective dates of publication are 8 December 2010, 3 March 2010, 19 January 2011, 12 January 2011, 5 January 2011, 15 December 2010, 22 December 2010, and 26 January 2011.

Bibliography

1. Kazuhisa Ogawa, *Japan's War Capability,* Ascom, 2005.
2. Nobuhiko Ochiai, *Urgent warning — The Nation Will Die,* Shogakukan, 2003.
3. Nobuhiko Ochiai, *The Birth of a Nation with the Best Information Strategy,* Shogakukan, 2007.
4. Buntaro Kuroi (editor), *The Total Picture of the American Information Agency,* JAPAN MILITARY REVIEW, July 2006 supplemental issue, *World Intelligence,* Vol. 1, JAPAN MILITARY REVIEW, 2006.
5. Masaru Kotani (editor), *World Intelligence — Deciphering the information wars of the 21st century,* PHP Institute, 2007.
6. Eisuke Sakakibara, *Japan Will Fall,* Asahi Shimbun, 2007.
7. Toshiyuki Shikata, *Japan Cannot Survive in this Situation,* PHP Institute, 2007.
8. Defense Research Center (compilation), *International Military Data 2008–2009,* ASAGUMO NEWS, 2008.

9. Gen Nakatani, *The Truth About the Ministry of Defense No One Was Able to Write About,* Gentosha Literary Publication, 2008.
10. Ministry of Defense (compilation), *2010 Edition: Japan's Defense — Defense White Paper,* GYOSEI, 2010. Available at http://www.clearing.mod.go.jp/hakusho_ data/2010/2010/figindex.html (20110528)

Conclusion

The Road to Achieving the Super Effect

Akira Ishikawa and Tetsuro Saisho

Thanks to the efforts of eight academic and business writers, we were able to collect around fifty super effects cases in the areas of new services and new businesses.

These cases were either categorized under those that featured technical evaluations made by academic researchers or those that featured business evaluations by front-line business professionals. Both of these types of cases highlighted sharp analyses and appreciation of the super effects from each of their respective viewpoints. Prior to examining these cases, they were merely recognized as data that simply conveyed fluctuations, but through rigorous analysis, we were made aware of the details of the dramatic super effects and came to renew our understanding of the significant sizes of those effects as well.

The cases collected 4 in this book showed the super effects implemented by organizations that realized objectives or enhanced functionalities in terms of targets aimed or scales of effect achieved, such as time reductions, speed increases, expansion of capability, improvement in responsiveness, improvement in the level of precision, increases in effectiveness and efficiency, changes of size, cost reductions, sales expansions, and quantitative increases and expansion in economic scale.

Furthermore, because it appears that we can use both the positive and negative aspects of the super effects in each case as keys to further our understanding, we would like to offer a summary of their characteristics and results.

Section 1: The Subjects and Scales of the Super Effects

With regard to the subjects of the super effects, there are a broad range of types, ranging from those that cover the international domain that feature observations of the world and the topic of international cooperation made from a macroscopic perspective, a single company observed from a microscopic perspective, and even those that cover not only large enterprises but also independent, small and intermediate businesses and venture corporations.

For example, when observing from a macroscopic standpoint, the case of the Intelligence Olympics presented by Akira Ishikawa is an event that aims to achieve an awareness for planethood and global security, and a qualitative progress in the standards of living, and while the context of the event is set in the spirit of international competition, the subject is one that is oriented toward global reform and renewal.

In addition, the case of the Earth Policy Institute presented by Toru Fujii relates to the design of geothermal power generation on a global scale with a plan to allocate 200,000 MW (megawatts) for the world's geothermal electric power generation. If we take into consideration the fact that there already are 24 countries carrying out geothermal power generation now, and that geothermal power generation makes up for more than 15% of all electric power consumed in five of those countries, we can clearly understand that this is a global-scale enterprise. Moreover, this enterprise is situated in the wider context of the good news that an amount of energy 50,000 times as much energy as found in all the world is deposited within 10km of the earth's crust.

Although the scope is a little smaller, the case of the smart meter project presented by Kazuo Matsude calls for improving the precision of measuring changes in electricity consumption in the daytime, instead of merely producing monthly totals for each household. In the area of the development of next-generation watt-hour meters, as evidenced by the fact that *Toko Electric Corporation*, *Toshiba*, *Fuji Electric*, and *GE* have joined forces to launch a joint venture, in addition to seeing *Google* carry out cooperative initiatives in Germany, we can understand that this area's international development is being rapidly pushed forward.

In the future, in preparation for an unforeseen development such as the Tohoku Earthquake and Tsunami Disaster that struck on March 11, 2011, and to plan for overall risk reduction by combining the smart meter with the conventional demand-response system, it is necessary to have such projects make progress on a global scale.

From the standpoint of international affairs, the case presented by Akira Ishikawa on dramatic changes in international competitiveness is interesting. While modern Japanese society, as observed from the perspective of the four major classifications of economic conditions, government efficiency, business efficiency, and infrastructure, manages to maintain its ranking within the top 10, in sixth place in terms of infrastructure, the nation ranks 34th in terms of government efficiency.

In particular, Japan must pay attention to the fact that the nation was judged lowly in a report on gender inequalities, ranking in 75th place among 134 countries. However, it is mentioned that this ranking was further downgraded to 101st place after receiving criticism from domestic women's groups. Rather than a positive super effect, this precedent cannot but be described as a typical case of a negative super effect. It is important to take precautions so as not to overlook the negative effects of super effects, as opposed to insisting on paying attention to only their positive effects. On the other hand, with regard to cases covering the single company or independent small-and intermediate-sized businesses, rather than just large enterprises, as observed from the micro-perspective, in terms of the sheer number of cases, this type has the most. It is these kinds of companies that are supporting the fundamental technologies of Japan, a techno-nation.

For example, the case of the cleaning operations presented by Tetsuro Saisho shows how *Gunkyo Factory*, a medium-sized company located in a local city, succeeded in reducing the cleaning operation cost by washing filters so as to make them as good as new for reuse. In so doing, the company brought down the cost of purchasing and disposing filters from 16,500 yen to a low cost of 5,900 yen. This is a super effect indeed.

Also mentioned is the super effect achieved as a capability that was necessary for *Mitsui* Warehouse to augment its main business of warehousing with a cargo business. This case shows how the company, in running its cargo business, was able to lower the cargo transportation cost from the 17,000 yen to 20,000 yen range to the surprisingly low 6,800 yen to 8,000 yen range by specializing in small-lot cargo and carrying out cargo consolidation *via* providing regular transport service and loading small quantities of freight from multiple clients.

Section 2: Targets and Functions of Super Effects

In modern US-style capitalist societies, where finance, accounting, and other monetary affairs play a central role, the key purpose of the corporation is to maximize revenues for sales and profitability while minimizing expenditures.

In a wider sense, since these aspects are also related to time reductions, improvements in precision and responsiveness, increases in effectiveness and efficiency, it is possible to grasp them as being integral to management functions.

Sales expansion and cost reduction

With regard to the typical case of sales expansion, Hiromichi Yasuoka took up the case of *Rakuten*, for example, to show how the company achieved a twenty-fold increase in its sales by entering into finance-related businesses, offering points programs for electronic money, an Internet bank, an online securities service, sales on credit, and a credit card. In addition, the author points out as a detail that cannot be overlooked the scale effect the company achieved by entering into over 70 business partnerships.

On a separate front, Atsushi Tsujimoto discovered that *CECIL McBEE*'s product development model fosters a "nesting" situation: a structure that fosters an interrelationship between the "act of observation" that supports product development and the "act of observation" made by the salesperson, while also enabling the former's achievement to lead to the success of the latter. In addition, the author introduces how a super-effect expansion in sales was brought about *via* connecting the organization with the "emotional" world of customers.

Furthermore, with regard to *P&G*'s superb marketing performance, Tsujimoto explains how the company has achieved a super effect with the aid of the Hierarchal Autonomous Communication System (HACS). The HACS model that played an important role here refers to a composite model in which a contemplating subject (such as a consumer monitor) and a subject who observes this contemplating subject are united to draw out valuable information (such as a product development manager who conducts hearings).

Hiromichi Yasuoka gives high acclaim for the activities of *Yahoo!* JAPAN, focusing attention on the company's ability to improve convenience through sharing IDs by leveraging the OpenID standard, along with praising the much wider concept of the company's economic results. In addition, he points out not only the increase in revenue achieved *via* electronic monies such as *Suica*, but also the drastic cost reduction achieved in the process, introducing this as a super spending effect.

By contrast, with regard to cost reductions, Akira Ishikawa introduced an epoch-making case that illustrated how a particular development of an artificial nano satellite helped lower its cost from 50 billion yen to 100 million yen, shrinking, in effect, the cost to a whopping 1/500th of what it originally was. This case has also seen the time of delivery reduced to realize completion of the project in one and a half years, boosting the cost reduction effect further.

Time reduction

Examining the impact of another function — time reduction — Kazuo Matsude raises the great achievement made in reducing the length of time it took to cover the distance between Nishikagoshima (currently Kagoshima Chuo) to Hakata with the conventional limited express bullet train, the *Super Ariake*. Namely, the reduction achieved was from 246 minutes to a blazing 79 minutes. This was made possible by opening traffic to the Kyushu Shinkansen and Matsude emphasized how this proved instrumental for not only achieving a time reduction, but also for realizing cross-border marketing.

Akira Ishikawa raised the case of *Eisai* Distribution to introduce how inspection time was shortened to 1/10th of what it normally was by taking advantage of RFIDs. Furthermore, he introduced a time-cost saving achieved through using *V-CUBE* technology to carry out web conferences and teleconferences, a saving that was also accompanied by a dramatic reduction in the actual cost. In one example, we saw how the use of *V-CUBE* technology helped a certain company in one quarter to drastically reduce travelling times from 162 hours to 34 hours, the time required for branch manager meetings from 102 hours to 34 hours, and in the case of the time spent on visiting the president of the company, the reduction achieved was from 60 hours to a surprising 0 hours.

Tetsuro Saisho raised the case of *Hikari Kogyo* (currently *Art-Hikari*) and explained how the company realized a drastic reduction in processing time, along with dramatically improving stacking capability, in the field of steel plate processing. Namely, the company applied seam welding to make it possible to support the processing of not only two stacks but four stacks as well, while boosting the processing speed five-fold from the 0.7–2m/minute range to the 0.7–10m/minute range.

In addition, Saisho also introduced the case of *Japan Highcomm* to illustrate how the company joined forces with a firm specializing in casting molds and a lost wax maker to develop a cast mold drying microwave dryer, applying the company's proprietary food processing microwave technology to reduce drying and thawing turnaround times to one-tenth as much.

Improvement of precision and sensitivity

Junpei Nakagawa, on the other hand, introduced the case of *Grace* that examined the re-visitation rates of customers who followed the company's *Twitter* account, and seeing that this rate exceeded 50%, he went on to show what role *Twitter* played in improving the retention ratio. Digging deeper with his investigation, Nakagawa also discovered that *Twitter* contributed to an increase of 1,600,000 cell phone contracts a year for *Softbank Mobile*, prompting him to commend how remarkably high *Twitter*'s levels of reliability and conveniences are.

In addition, Nakagawa reported on how *Ryohin Keikaku* went ahead with a new participatory approach to development that involves customers to achieve a thirteen-fold increase in customer registrations with exclusive limited events. This is analogous to *Tokyu Hands* developing new business activities and achieving a fifteen-fold increase in followers.

Furthermore, he introduced how *Dell* produced an information diffusion effect through skillful implementations of limited discounts to realize a 150-fold increase in the number of its followers, as opposed to interested parties. He describes this as a result that aimed for an indirect effect as opposed to a direct one.

On the other hand, Kazuo Matsude took a look at *Toray*'s seawater desalination business made possible *via* the process of reverse osmosis through a membrane and noted that this domain is ultra-structural, using an aperture that is, by a large margin, less than one nanometer. For this reason, he pointed out the need for extremely fine-tuned precision in this domain. And he also pointed out that most large-sized industrial plants have now come to use the reverse osmosis technique as its cost has markedly reduced relative to the conventional evaporation method.

Akira Ishikawa reported on how the course of the development of the carbon nanotube came about serendipitously, while showing that this material, when compared to copper, has a high current density tolerance of more than 1,000 times, along with a high heat conduction characteristic of 10 times as much. Furthermore, when comparing to aluminum, the material is half as light, and in comparison with steel, its tensile strength is 20 times more. With regard to its tensile strength when pulling in the direction of its fibers, Ishikawa reports that it even surpasses diamond. The carbon nanotube is anticipated to serve as raw material for the ropes to be used in space elevators in future.

Speed boost

An increase in speed has an inverse relationship with time reduction. Takashi Yonezawa introduced the case of *Ishida*, showing how the company developed a computer scale that realizes superior speed and precision through the application of the multihead weigher. This is a case study of the successful development of a computer scale that made it possible to make measurements 200 times faster than before, moreover with a margin of error of around 19, even when individual particle sizes varied. For this reason, this scale has been appraised as an international standard, accounting for a domestic share of 80% and a global share of 70%.

Additionally, Yonezawa reported on how *Tsubakimoto Chain* applied an innovative technology to achieve speed boosts 10 times greater while raising the durability level 10 times greater in the domain of table lifts (Zip Chain Lifter), used for the vertical movement of people and objects.

Furthermore, Yonezawa described how *Anzai* Manufacturing developed the *RSC-1800*, a sorter that can separate non-conforming articles. With a capability of processing polished rice that weighs 0.02g per rice grain at a throughput speed of 18ton/hour, this sorter can function at the staggering speed of approximately 1,000 grains per second per channel.

On a different note, Yonezawa touched on the topic of the *IBM ILOG CPLEX Solver* and explained that the computational performance had improved 800 times in comparison with the *CPLEX V1.0*, which was introduced in 1988, and the *CPLEX V8.0*, which was introduced in 2002.

However, he went on to emphasize that since algorithmic improvements led to the speeding up of the throughput speed by 2,360 times as much, taking this effect, along with the aforementioned one, into account, the *IBM ILOG CPLEX*'s performance enhancement has, in fact, achieved a 1,900,000-fold improvement.

Boosts in effectiveness and efficiency

Increases in effectiveness and efficiency are often compared with cost (expenses) *vs*. effect (efficiency) analyses. While the cost is reduced, if there is increased effectiveness, we can kill two birds with one stone.

Toru Fujii raised the case of *Mitsubishi Heavy Industries* to show how the company developed an infrastructure for electric buses that run on replaceable batteries to help advance the realization of a society with a low carbon footprint, pointing out that this had led to the amount of emission per passenger being dramatically reduced to one-eight its former amount.

Kazuo Matsude looked into the case of Ireland's *Ryanair*, a low-cost carrier (LCC) to find out that the fare for flying from London to Oslo was a surprising 945 yen, markedly standing out in contrast to *British Airways*, which charged 9,693 yen, *Scandinavian Airlines*, which charged 9,383 yen, and *Norway Airlines*, which charged 6,210 yen. This cost reduction contributes toward globalization of regional movements for tourists, leading to increases in efficiency and frequency of travel.

Junpei Nakagawa investigated *Yamada Denki*, the company based in Gunma that aimed to develop its business along roadsides, and reported that it had overcome the impact of the 2011 Tohoku Earthquake and Tsunami Disaster, with their net income up 68.5% from a year ago, their best gain to date.

Nakagawa goes on to elucidate that one of the reasons for this outcome was the fact that commentators holding the key to spreading information about the products with their assessments were able to receive information on the arrival of goods through *Yamada Denki*'s tweets and then post comments such as, "I was able to buy a great product", thereby increasing the number of customers among their followers. This in turn led to the increase in the number of new followers for the store's *Twitter* account in real-time. The author describes this spill-over effect as synergistic marketing.

Toru Fujii, on the other hand, introduced the case of *Hitachi Construction Machinery* in which he showed how the company, with the help of the domestic credit system, realized a 64% reduction in CO_2 emissions compared to its conventional level by utilizing the electrically-driven hydraulic shovel for small-and intermediate-sized businesses.

Furthermore, Tetsuro Saisho introduced the case that saw *Panasonic* achieving an electricity consumption level of 15% without reducing its amount of production. This was made possible through a setup that entailed the installation of sensors measuring electricity usage throughout *Panasonic*'s plants to gather data that could help eliminate wasteful uses of electricity. Since this setup can be adapted by not only *Panasonic* and other large companies as their in-house countermeasure, but also more widely by competitors and even manufacturing firms of other industries, in addition to smaller, high-rated firms, its wider application throughout modern society is certainly to be hoped for.

Capability augmentation

As an exemplary case for illustrating ways that lead to the augmentation of capabilities in conjunction with increases in effectiveness and efficiency, we cannot go wrong by raising the case example of the *Tokyo Stock Exchange* as put forth by Tetsuro Saisho.

Saisho reports that the company managed to augment its capabilities from 400 to 600 times as much as the capabilities of its former system by speeding up the same stock exchange's order response times and information transmissions. In addition to carrying out rapid time reduction, this new system, named *arrowhead* [sic], takes both scalability and reliability into consideration.

Furthermore, Saisho also mentions that by pushing forward with a joint research collaboration with Tokyo University and *Sharp*, the company is engaging in the development of a photovoltaic power generation facility capable of producing one million kilowatts of energy, the world's largest output (which is 10 times greater than the conventional output). To this end, the project members will be carrying out a demonstration in Saudi Arabia by 2014, and depending on the favorable playing out of the "Special Measures Law for Renewable Energy" currently being advanced by Japan's *Diet*, it is thought that photovoltaic power generation will spread explosively throughout the entire nation, making it likely that a super effect will appear in this domain as well.

Kazuo Matsude raised the case of *Hitachi* that illustrated how the company reduced on the time required to start up terminals from two to three minutes to around 10 seconds by introducing virtual OS-based thin clients. He went on to point out how dramatically well this time reduction has contributed toward the company's productivity and capacity expansion.

Toru Fujii surveyed smart cities and smart communities and mentioned that its market has grown in 2020 to a worth of 180 trillion yen within a decade, quadrupling in scale. He goes on to mention that there are currently 300 to 400 smart city projects underway in the world.

His assertion is that the period from 2010 through 2020 promises to be rife with business opportunities for Japan, since even though at the outset North America and Europe will be the main drivers of growth, after 2015, China will supersede them in this role, and after 2020, smart cities and communities stand to experience growth in India.

Improvement in responsiveness

Along with capability, an improvement in responsiveness, viewed as a functional improvement, cannot be ignored. For example, Junpei Nakagawa went on to apply the same analysis he made for the tweeting activities of the aforementioned *Grace* and *Yamada Denki* to the shopping district in Suginami-ku to find out that individual stores themselves in the Koenjilook Shopping Street played the role of organizers, and to that end, made use of *Twitter* to revitalize the area, tweeting the new arrival of goods. This initiative resulted in 6,000 followers. In other words, what this case showed was that it wasn't commentators making recommendations that contributed toward revitalization, but their followers, who improved responsiveness to sales by way of receiving those recommendations from the leaders.

Amid a situation where we are seeing customer traffic flow from small boutique stores to large-scale chain stores in not only local cities but within the Tokyo Metropolitan area as well, *Twitter* has become one means to check this tide.

On the other hand, Kazuo Matsude observed that *Amazon.com*, headquartered in Seattle, Washington, U.S.A., had gross sales of 34,204,000,000 dollars, making it the world's largest company in those terms, and maintains that the reason behind its popularity lies in its effective data mining capability. This has been said to provide an experience that is not unlike having an exclusive personal librarian.

The company not only retains the book purchase histories of customers for the past eight years, but also carefully saves the history of the books they checked but didn't purchase, so as to present detailed recommendations of various books based on the trends evidenced through such data. The more such information accumulates, the more fine-tuned the recommendations can become.

In other words, by carrying out such data mining and improving each customer's sensitivity to their own tastes, *Amazon* is accumulating value-added sales.

Size change

When speaking of sizes, we can talk about sizes in a physical sense or sizes as perceived by the mind or feelings.

Atsushi Tsujimoto took note of this point and cited the case of the brand *Francfranc* as an exemplary study in how the company achieved success by having its staff members recognize everyday situations and their characteristic signals not only in a visual sense, but also in a both physical and emotional sense. In particular, the success of the company was achieved precisely through reflecting such observations in its product development projects. This line of thinking is grounded in the product developer's drive to develop products associated with moving mental imageries (arising from everyday life) that like-minded consumers can sympathize with and relate to.

According to the author, thanks to this line of thinking, the company's consolidated sales in domestic and foreign markets as of the January 2011 period became 33 billion yen, causing *Francfranc*'s consolidated sales to account for 86% of sales. In essence, this was a case that illustrated a change not in physical size, but in emotional size.

Akira Ishikawa cast a spotlight on the dramatic potential of the e-learning market and reported that Japan's e-learning utilization factor was 3.1%, according to the 2005 edition of the white paper on telecommunications, but that in the US, this was already considerably more — at 24%, while in Korea it was even more at 25%.

And drawing from Terunobu Kinoshita's work, "The Age of the Digital Natives," Ishikawa pointed out that, while the market size has not reached even 10%, grass-root level e-learning is catching on in Japan, with the market for PC and Internet-based e-learning aimed at individuals seeing a drastic 25.7% increase and the market for e-learning programs designed for game consoles seeing an amazing 41.6% rise.

This is a market size expansion, showing the potential for enlarging to a significant scale, just as in the case of Korea, where micro-effects at the personal level worked together with the macro effects triggered by the government.

On the other hand, Hiromichi Yasuoka brought to our attention the dramatic increase in demand for consulting *via* information dispatches, underscoring the need in the field of consultancy to have its practitioners embrace "an all-encompassing perspective and convey the true nature of subject matters by staying one step ahead of their clients (not two)".

Yasuoka discusses how he has worked to improve the quality of information dispatches and the quality of his expertise to ultimately contribute to the consulting firm to which he belongs a total number of orders that exceeded the amount of 100 million yen on four occasions. Furthermore, he emphasized that he had indeed, relatively speaking, prompted a super effect with such a contribution, since the number of consultants achieving such a performance only amounted to 1/10th of the managerial class, which, according to a simple calculation, makes the figure 1/10,000th.

This information dispatch technique he advocates is what is known as mediative profiteering, or the tactic of introducing a problem to be rewarded for solving it. It is believed to have expanded the size of organizations and contributed toward augmenting their capabilities and competitive edge.

Section 3: Other Considerations

Discovery of a raw material itself

In the course of realizing a super effect, we cannot overlook the importance of efforts put into discovering raw materials — those materials that seed purpose and function.

Kazuo Matsude reported that Professor Makoto Watanabe and others from Tsukuba University announced their discovery of *Aurantiochytrium*, a promising algal strain capable of harvesting biofuel. The announcement was made on December 14, 2010 at *The 1st Asia-Oceania Innovation Summit*.

The production efficiency of this algal fuel is reported to be 10 times that of *Botryococcus*, which had the best production efficiency until the discovery was made.

Furthermore, according to provisional calculations, the cost of extracting fuel from *Botryococcus*, under existing circumstances, will end up amounting to 800 yen per liter at the most, whereas in the case of *Aurantiochytrium*, with its large-scale cultivation, the cost of producing fuel can be brought down to around 50 yen per liter, a study group surmises. If that price level comes to fruition, the raw material will become viable as a commercial product and will be catapulted into the limelight.

The negative effects of super effects

The most remarkable example of the negative side to a super effect is the weakening of Japan's competitiveness as seen from the perspective of the Balance of Power theory.

Namely, this example is the inexcusable reality that, relative to Japan's neighbors in Asia, North Korea, China, Russia, India, and Pakistan, Japan possesses no nuclear weapons, with the disparity between military forces standing at 20%, according to the 2010 edition of the Defense White Paper.

According to Organski and Kugler's analysis of the Franco-Prussian War, Russo-Japanese War, World War I, and World War II, which looked at the changes in the levels of national power over a 20-year period prior to the outbreak of each of those wars, the likelihood of a nation with more than a 20% power disparity going to war was more than 50%.

To find a way out of such a critical situation, Akira Ishikawa, like some commentators on military affairs, advocates the establishment of the Aerospace Self-Defense Force, which can make modern warfare impossible to execute. This force will, in effect, aim to make use of outer space.

On a separate note not covered in this book, according to Ishikawa, Japan's legal administration is in a state of crisis, with its number of legal experts paling in comparison to the numbers seen in the UK, the United States, France, and Germany. For this reason, he says the occurrences of cases involving miscarriages of justice are unavoidable.

To be specific, according to the 2004 Court of Appeals Data Book, the number of judges in Japan was 2,385, whereas this number was 31,275 in the US; 20,901 in Germany; 5,257 in France, and 3,803 in the UK. In addition, in terms of the number of judges per 100,000 people, Japan had 1.87 judges, whereas the US had 10.85, Germany had 25.33, France had 8.78, and the UK had 7.25, making the difference, for the most part, off by an entire digit. With such statistics, the reality proves to be untenable.

Bibliography

1. Terunobu Kinoshita, *The Age of the Digital Natives — A portrait of the youths changing the next era* (Seikatsujin Shinsho), Toyo Keizai Shinposha, 2009.
2. Ministry of Defense, *Defense of Japan 2010 — Defense White Paper*, GYOSEI, 2010.

About the Editors and Authors

Editors

Akira Ishikawa

Professor Emeritus at Aoyama Gakuin University, Dr. Ishikawa is also a Senior Research Fellow with the University of Texas ICC Institute. He received his Ph.D. in Management from the Graduate School of the University of Texas and completed postdoctoral studies at MIT.

After successively holding posts as Assistant Professor at the Business Management Graduate School of New York University, as Associate and Full Professor at Rutgers University's Graduate School of Management, and as Affiliate Professor at the Graduate School of the University of Hawaii, he went on in 1984 to serve as Professor at the Graduate School of International Politics, Economics and Communications of Aoyama Gakuin University, and in 2001 as Director of this school. In April 2003, he was appointed as Professor Emeritus of Aoyama Gakuin University.

Until recently, he was also engaged as a teacher by the graduate schools of Tsukuba University, Temple University, and the International University of Japan. His areas of expertise lie in Management Science, Management Accounting, Crisis Management, Research and Development, Accounting, and Finance for content businesses.

His other chief works include *Strategic Budgeting Control* (sole author, Doubunkan Shuppan, 1993), *Approach to Acquiring and Modeling Intellectual Capital: Intellectual Capital Management (ICM) (*Aoyama Management Review No. 3, *Nikkei BP Kikaku* [currently *Nikkei BP Consulting*], pp. 24–33, 2003). He also has over 100 books to his credit as a collaborating author, editor or translator, and has presented more than 450 papers to date.

Tetsuro Saisho

Professor of Societal Informatics at the Faculty of Social and Information Studies of the Graduate Division of Gunma University, Dr. Saisho received his Ph.D. in Engineering from Chuo University. After graduating from the Department of Economics of Chuo University in March 1986, and holding a post at a financial institution, in April 2004, Dr. Saisho became Associate Professor of Economics at the Graduate Division of Economics of Kanto Gakuin University, and in April 2009, he served as Professor there for a period of four months before becoming incumbent Professor to date since April 2010.

His areas of expertise lie in Information Management, Information Systems, Information Security, and Management Strategy. At present, he also holds multiple posts as the Director of the Japan Society for Information and Management, Director of the Japan Society of Security Management, Director of the Japan Society for Systems Audits, Councillor of the Japan Information-Culturology Society and Chairperson of its Public Relations Committee, Member of the Editorial Committee of the Bulletin of the Transdisciplinary Federation of Science and Technology, Member of the Editorial Committee of the Bulletin of the Japan Academy of Business Administration, Member of the Research-activities committee of The Society of Social-Informatics.

His other principal works include, *Information Security Management of The Modern Organization: The Strategy of Information Security Management, Introduction, Decision, Employment* (author, Hakuto-Shobo Publishing Company, 2012), *The Innovation System of China and Vietnam: The Innovation Creation Strategy by The Industrial Clusters* (author, Hakuto-Shobo Publishing Company, 2011), *Information Strategies of the Modern Enterprise and Corporate Transformation* (author, Hakuto-Shobo Publishing Company, 2009), *Introduction and Development of Information Security Management* (author, Kanto Gakuin University Press, 2006), and *Modern Management in Japan and China* (co-author, Yachiyo Shuppan, 2009), and *Creative Marketing for New Product and New Business Development* (co-author, World Scientific Publishing Company, 2008).

Authors

Atsushi Tsujimoto

Honorary Researcher of The University of Tokyo Interfaculty Initiative in Information Studies — Graduate School of Interdisciplinary Information Studies, Professor Tsujimoto received his Ph.D. in March 2006 from the Doctoral Program of the Graduate School of Humanities and Sociology of Tokyo University. At present, he concurrently holds the post of lecturer at Meiji University. His areas of expertise lie in Organizational Theory and Risk Management Theory and one of his major works is *Creative Marketing for New Product and New Business Development* (Co-editor, Seisansei Shuppan, 2006).

Junpei Nakagawa

Associate Professor at the Business Administration Department of Komazawa University, Professor Nakagawa received his Ph.D. from the Economics Program of the Graduate Division of Tokyo University. His areas of expertise lie in Management Theory and the Theory of the firm. His chief works are *Theory of Business Administration for Problem Solving* (sole author, SENBUNDO, 2007) and *Iwanami Textbook — Contemporary Economics* (Collaborating author, Iwanami Shoten, 2008).

Toru Fujii

Mr. Fujii received his Ph.D. in Integrated Policy Studies from Chuo University and is now with *Hitachi*, serving as a Deputy General Manager of the company's Solution Sales/Marketing Division and Business Planning Department.

After graduating in March 1990 from the Faculty of Social Sciences of Hosei University, he joined *Hitachi*, Ltd. and concurrently holds today the post of Director of The Japan Society for Management Information, Director of the Japan Academic Society of Hospitality Management, Member of the Bulletin Editorial Committee of the Transdisciplinary Federation of Science and Technology, and Collaborating Researcher at the Faculty of Social and Information Studies of Gunma University. His area of expertise lies in Service Science and Strategic Management of

Technology, and one of his major works is *Smart-Infra Strategy* (author, V2-Solution, 2012).

Kazuo Matsude

Mr. Matsude graduated from the Faculty of Law of The University of Tokyo in March 1986 prior to joining a major financial institution, where he engages in international business. He earned his MBA from Yale University and is a member of The Japan Society for Management Information and Japan Association for Management Systems. At present, he also concurrently holds the post of Collaborating Researcher at the Faculty of Social and Information Studies of Gunma University. His areas of expertise lie in the Theory of Information Systems Management and International Finance Theory.

Hiromichi Yasuoka

Dr. Yasuoka is with Nomura Research Institute, Ltd., serving as a Senior Consultant in the firm's Consulting Sector. He received his Ph.D. in System Design and Management upon completing this field's doctoral course from the Graduate School of Keio University. At present, he also concurrently holds the posts of part-time lecturer at the Faculty of Business Administration in Komazawa University and at the Faculty of Engineering in Daiichi Institute of Technology. His area of expertise lies in Business Strategic Planning and developing startup ventures that offer loyalty points programs, electronic money, ID, and payments applications for various fields, ranging from Information and Communication Technology to Financial Services.

Takashi Yonezawa

Equipped with a Bachelor of Science in Physics from Kyoto University, Mr. Yonezawa joined *IBM* Japan in 1989 and went on to become an IT specialist with *IBM* Business Consulting Services KK. He currently helms the company's Business Analysis Optimization (BAO) Division. His areas of expertise lie in the actual application of optimization technologies for businesses and in the optimization of system architectures, particularly in the Supply Chain Management (SCM) domain.

Afterword

Herewith, we publish this book titled *Corporate Strategy for Dramatic Productivity Surge* as a work of collaboration between eight authors. In this book, four academicians and four businessmen each contributed case examples of super effects brought about by new, cutting-edge products — effects we termed *ketachigai* in Japanese, a word used to describe both positive and negative differentials that are off by an entire digit.

The four academic authors each are representatives of their respective generations and carry out state-of-the art research in the theory of business administration. While I also belong to this academic camp as one of the editors of this book, I have experience in actual business management as well.

The universities I and the other academicians are affiliated with are all located within the Kanto region and have departments and graduate schools specializing in various areas — such as management and social information studies — where analyses and evaluations are carried out on the latest case studies as seen from the expert viewpoint of each researcher.

In addition, each of the four authors from the business world are key persons who belong to major, prominent firms that represent Japan to the world, and play an active role as trailblazers in their respective fields. Furthermore, the types of business to which these business professionals belong extend widely, ranging from general electrical equipment manufacturers to think tanks, where analyses and evaluations are carried out on the latest case studies as seen from the seasoned viewpoint of each businessman.

In this way, the most unique point about this book is the fact that both academicians and businessmen were able to present assessments on business models related to new products and new businesses from all conceivable domains without being fixated on preconceived notions about specific fields, business sectors, industry types or the scale of enterprises.

With regard to the businessmen's cases, I would like to emphasize in particular that they were in no way appeals for products or services that their respective companies specialize in. Instead, they are the fruits of examining precedents from their own unbiased, objective perspectives and investigating the substance of rival firms, competitors, and organizations that have competing business models, adding highly insightful evaluations on the nature and content of those precedents in the process. In other words, what is noteworthy here is that it is businessmen themselves who are taking up cases of their rival firms and other competing companies to discuss and present business models related to new products and new businesses for the purpose of evaluating dramatic super effects.

Incidentally, seven years have already passed since I had found my way into the academic community after going through more than 20 years of work experience in the business world. During these seven years, economic conditions have continued to be in a state of chaos, as they continue to be to this day, and amid such circumstances, many venture corporations have been born under the guidance of new business models. Meanwhile, many listed companies, smaller, high-rated firms, and even venture corporations are being cornered into closing their operations or becoming bankrupt.

For all existing companies (including large enterprises and smaller, high-rated firms) to grow permanently and realize dynamic development, the companies themselves must secure particular capabilities and develop strategies that respond to environmental variations surrounding them.

One such capability is the type that yields a super effect in the domain of new products and new businesses.

International competition as well as domestic competition related to R&D and business development and improvement are intensifying by the day, and dramatic results are being sought after from experts, regardless of whether their expertise stems from the fields of science and technology or the humanities and social sciences. If we neglect to attain these results, dropping out of the international competition, not to mention the domestic competition, is inevitable.

The necessity and significance is extremely high for collecting, analyzing and evaluating case studies of companies that produced super effects when they faced the challenges of the day and achieved outstanding results — that is to say, when they either sustained their results by lowering their costs to one-tenth as much or achieved a performance that was 10 times as much while maintaining their costs at their previous levels.

We are convinced that by collecting and performing deep analysis, and evaluating the case studies of such companies, we can offer one of the key resources that can aid large enterprises, along with smaller, high-rated firms, in their attempt to reinforce their competitive edge, paving the way in particular toward restoring their global competitiveness.

Since all business operations are involved in these cases, not to mention the people directly in charge of R&D and business development, it is necessary to push ahead with those operations every day with a top-of-mind awareness of the cases. In particular, from now on, we believe that they will serve as essential references for undergraduate and graduate students, interns, researchers and others in academia, not to mention the business people or executives related to the businesses covered by the cases.

On a separate note, due to space limitations, I feel that we were not able to fully cover overall details related to fields, industry types, and business conditions associated with super effects, and for this reason we were only able to offer our viewpoints on a case-by-case basis, making our observations which lack in thoroughness in some respects. However, I would like to examine these matters as future research themes, focusing my lens of inquiry on those super effects related to new products and new businesses found in the post-industrial information society. Of course, needless to say, the final responsibility for the contents of each case study in this book lies with each of their respective authors.

Finally, to you, dear reader, I would like to express my appreciation for your interest. There is nothing that will make us more grateful than having you, through this book, take interest, however little that may be, in the nature of super effects found in the domain of new products and new businesses and understand them, or come to appreciate what types of results are attained through super effects, what types of mechanisms or structures these super effects bear, and what types of forecasts can be made from them in the functional aspects that are directly related to productivity, such as time reductions, speed increases, capability augmentations, heightened responsiveness, heightened precision, increases in effectiveness and efficiency, changes in size, and discovery of advanced materials.

March 30, 2013.
Coeditor and author
Tetsuro Saisho

Index